Praise for

THE LAST SEASON

"A hell of a story, a tale of lost souls, and a one-of-a-kind look into a truly rarefied American subculture. I have a feeling *The Last Season* is going to be around for a long time, earning a place in every home library devoted to the California wilderness experience."

—Daniel Duane, author of *Caught Inside*

"Impeccably researched and compassionately told. . . . A compelling story of one man's passion and pain."

—Jennifer Jordan, author of *Savage Summit*

"A 'must-read' for any serious Sierraphile."

—Gene Rose, Sierra author, historian, and lecturer

"Very artfully written, *The Last Season* allows the reader a highly intimate—almost voyeuristic—insight into the life and mysterious disappearance of this enigmatic and, some would say, larger-than-life backcountry park ranger. I highly recommend it."

—Butch Farabee, author of *Death, Daring, and Disaster: Search and Rescue in the National Parks*

"A mesmerizing tale of one man's struggle for fidelity—to the woman with whom he joined his life, and to the wild world of which he was steward. Blehm sinks our teeth right into the marrow of that conflict."

—Amy Irvine McHarg, wilderness activist and author

"Eric Blehm's *The Last Season* will keep you reading into the night, and remain with you long after you have finished."

—Nora Gallagher, author of *Practicing Resurrection*

THE
LAST SEASON

ALSO BY ERIC BLEHM

*Agents of Change: The Story of
DC Shoes and Its Athletes*

P3: Pipes, Parks, and Powder

HarperCollins*Publishers*

THE
LAST SEASON

ERIC BLEHM

HarperCollins books may be purchased for educational, business, or sales promotional use. For information, please write: Special Markets Department, HarperCollins Publishers, 10 East 53rd Street, New York, NY 10022.

Title page photograph © 2006 by Richard Leversee.

FIRST EDITION

Designed by Renata Di Biase

Printed on acid-free paper

Library of Congress Cataloging-in-Publication Data
Blehm, Eric.
    The last season / Eric Blehm.—1st ed.
      p.   cm.
    ISBN-10: 0-06-058300-2
    ISBN-13: 978-0-06-058300-2
    1. Morgenson, James Randall, 1942–1996. Park Rangers—United States—Biography. 3. Sequoia National Park (Calif.) 4. Kings Canyon National Park (Calif.) I. Title.
    SB481.6.M67B58 2006
    333.78'3'092—dc22
    [B]   2005052606

06 07 08 09 10 RRD 10 9 8 7 6 5 4 3 2 1

For the unsung heroes of the
National Park Service

CONTENTS

THE
LAST SEASON

PROLOGUE

In the vast Sierra wilderness, far to the southward of the
famous Yosemite Valley, there is yet a grander valley of
the same kind. It is situated on the South Fork of the
Kings River, above the most extensive groves and forests
of the giant sequoia, and beneath the shadows of the
highest mountains in the range, where canyons are the
deepest and the snow-laden peaks are crowded most
closely together.

—John Muir, 1891

The 1996 season . . . could be written in the chronicles
of Sequoia/Kings Canyon National Parks as the one
season we hope never to have to repeat. The most
significant element in this history was the search for a
fellow park ranger and friend, Randy Morgenson.

—Cindy Purcell, Kings Canyon subdistrict ranger, 1996

IF CHINA HAD BEEN ENDOWED with a well-placed mountain range
like that of the southern Sierra Nevada, its Great Wall would not have
been necessary.

The Sierra's formidable granite spires, snowy white most of the year,
parallel the Pacific Ocean, north to south for more than 400 inland
miles. In the southern part of the range, the ramparts are highest and

steepest, and a double crest—like a castle's inner and outer walls—is at once daunting and seemingly impassable. Between these walls of jagged peaks runs the mighty Kern River, an icy torrent twisting and cascading southward through a maze of lesser peaks and forbidding canyons to eventually irrigate the crops and orchards of California's San Joaquin Valley.

Though a few hardy souls cross these mountains in winter, most wait until the snow melts, when access to the high country can be attained via a network of routes that evolved over the centuries from threadlike, barely perceptible game trails. These ancient animal paths were widened slightly by the native populations, who used them as trade routes between the coast and inland valleys and deserts. They were later trampled by herds of domesticated sheep, and eventually blasted by dynamite, graded, and manicured with pick and shovel for recreational purposes by the Civilian Conservation Corps during the Great Depression.

There are few blacktop passageways running east to west in the entire Sierra range, and none running north to south for any distance. South of Yosemite National Park is a conspicuous absence of blacktop for over 200 miles. This wilderness area is concentrated within the boundaries of Sequoia and Kings Canyon National Parks—two adjacent parks managed as one 860,000-acre unit. According to government records, Sequoia was founded on September 25, 1890, and is the second-oldest national park, after Yellowstone. Kings Canyon was originally founded on October 1, 1890, as General Grant National Park—the country's third national park. It was renamed Kings Canyon on March 4, 1940. Some 70 percent of Sequoia's 402,510 acres is designated wilderness and nearly 98 percent of Kings Canyon's 461,901 acres is wilderness. The combined wilderness areas—essentially roadless backcountry—covers roughly 1,350 square miles.

Here the most traveled human thoroughfare is the John Muir Trail. Jokingly referred to as a freeway, it is rarely wide enough for two backpackers to walk shoulder to shoulder. The trail was conceived of by Theodore Solomons, who in 1884 dreamed of a remote trail atop the crest of the High Sierra. Construction began in 1892, and in 1938 the completed trail started at an elevation of 4,000 feet in Yosemite Valley and traveled 211 miles south over ten mountain passes before ending at the 14,495-foot-high summit of Mount Whitney. Overlapping the 2,650-mile Pacific Crest Trail, which runs between the Canadian border and Mexico, the John Muir Trail is the highest, remotest, and most grueling segment of the Pacific Crest Trail.

More than 800 miles of trails wind their way up into the high country and are accessed by more than thirty trailheads on the east and west sides of the range. The western approaches, in contrast to the eastern ones, are gentler in slope—escalators versus elevators. Almost all trails lead eventually to the John Muir Trail. It is estimated that 99 percent of the visitors to the parks' backcountry stay on these designated tracks, which represent less than 1 percent of the parks' wilderness acreage. True to the idea of wilderness, 99 percent of the parks' backcountry is raw and wild. A craggy, high-altitude desert of granite and metamorphic rock dominates the country. But dotting the arid landscape of serrated ridgelines and glacial sculpted domes are remnants of the last Ice Age, or at least the last winter: striking sapphire blue lakes, ribboned inlets and outlets become creeks snaking across arctic-like tundra, giving drink to vibrant brushstrokes of meadows and forests, while swatches of green erupt like oases from the volcanic and glacially formed grayness. The contrast softens the hard, rocky vistas and coaxes ecosystems to take up residence amid the harshness of it all.

There are no year-round residents, at least of the two-legged variety. The only structures are summer ranger stations, many of which double as snow survey cabins in winter, and a handful of historical trapper cabins and mines that are slowly being reclaimed by the wilderness. The stations are located every 20 miles or so along the major trails and are inhabited from June to October by seasonal backcountry rangers, men and women who have served for decades as quiet guardians of this national treasure and the travelers who pass through it. They are a special breed, these elite few—dedicated, fearless, and determined—and their reasons for seeking the splendor and isolation of wilderness are as varied as the geography they protect.

In the wilderness, life is reduced to its essentials: food, shelter, water. A person can lose himself here, both figuratively and literally. With very little effort, one can escape almost everything and everyone associated with civilization.

But the reflection in a clear mountain lake of one highly trained ranger serves as a reminder: What one cannot escape is one's self.

CHAPTER ONE

MISSING

I shall go on some last wilderness trip, to a place I have known and loved. I shall not return.
                    —*Everett Ruess, 1931*

The least I owe these mountains is a body.
                    —*Randy Morgenson, McClure Meadow, 1994*

THE BENCH LAKE RANGER STATION in Kings Canyon National Park was still in shadow when Randy Morgenson awoke on July 21, 1996. As the sun painted the craggy granite ridgelines surrounding this High Sierra basin, a hermit thrush broke the alpine silence, bringing to life the nearby creek that had muted into white noise over the course of the night.

A glance at his makeshift thermometer, a galvanized steel bucket filled with spring water, told him it hadn't dropped below freezing overnight. But it was still cold enough at 10,800 feet to warrant hovering close to the two-burner Coleman stove that was slow to boil a morning cup of coffee. If he had followed his normal routine, Randy had slept in the open, having spread out his sleeping bag on a gravelly flat spot speckled with black obsidian flakes a few steps from the outpost. Hardly the log cabin vision that the words "ranger station" evoke, the primitive residence was little more than a 12-by-15-foot canvas tent set up on a plywood platform. A few steel bear-proof storage lockers and a picnic table completed what was really a base camp from which to strike out into the roughly 50 square miles of wilderness that was Randy's patrol area.

Before, or more likely after, the hermit thrush's performance—assuming he followed his custom before a long hike—Randy ate a hearty "gut bomb" breakfast of thick buckwheat pancakes with slabs of butter and maple syrup. Then began the ritual of loading his Dana Design backpack for an extended patrol. Methodically, he stuffed his sleeping bag into the bottom, followed by a small dented pot—blackened on the bottom—that held a lightweight backpacker stove wedged in place by a sponge so it wouldn't rattle. A "bivy" sack was emergency shelter. A single 22-ounce fuel bottle, a beefed-up first aid kit, a headlamp, food—each item was a necessity with a preordained spot in his pack.

He locked his treasured camera equipment, six books, and a diary inside a heavy-duty "rat-proof" steel footlocker that was "pretty good at keeping rodents out too," he'd been known to say. His only source for contacting the outside world—a new Motorola HT1000 radio, along with freshly charged batteries—was zipped into the easily accessible uppermost compartment of his pack. This was the second radio he'd been issued that season; the first one had lasted only eight days before it stopped working on July 8. On July 10 he'd hiked over Pinchot Pass to the trail-crew camp at the White Fork of the Kings River, the location he'd arranged in advance with his supervisor if his radio conked out. A backcountry ranger named Rick Sanger had met him there with the replacement Motorola he now carried.

The least-used item in his pack was a Sequoia and Kings Canyon topographic map. He reportedly referenced it only while trying to orient lost or confused backpackers, or during a search-and-rescue operation. As longtime friend and former supervisor, retired Sierra Crest Subdistrict Ranger Alden Nash, says, "Randy knew the country better than the map did."

For nearly three decades, when someone went missing in Sequoia and Kings Canyon National Parks, standard operating procedure had included at least a radio call to Randy, the parks' most dependable source of high-country knowledge.

"Randy was so in sync with the mountains," says Nash, "that he could look at a missing person's last known whereabouts on a topographic map, consider the terrain and 'how it pulls at a person,' and make a judgment call with astounding results.

"One time, a Boy Scout hiking in the park got separated from his troop and couldn't be found before nightfall. Randy looked at a map for a few minutes, traced his thumb over a few lines, and then tapped his finger on a meadow. 'Go land a helicopter in that meadow tomorrow morning,' he said. 'That's where he'll be.'

"Sure enough, the Boy Scout came running out of the woods after the helicopter landed in that meadow. He'd taken a wrong turn at a confusing trail intersection and hadn't realized his mistake until it was almost dark and too late to retrace his footprints. The Scout was scared after a night alone, but he was fine.

"Randy," says Nash, "had figured that out by looking at a map. He told me where to go over the radio. John Muir himself couldn't have done that. But then, Muir didn't spend as much time in the Sierra as Randy."

A bold statement, but true. At 54, Randy had spent most of his life in the Sierra. This included twenty-eight full summers as a backcountry ranger and the better part of a dozen winters in the high country as a Nordic ski ranger, snow surveyor, and backcountry winter ranger. Add to that an enviable childhood spent growing up in Yosemite Valley—where his father worked for that park's benchmark concessionaire, Yosemite Park and Curry Company—and Randy had literally been bred for the storied life he would lead as a ranger.

His backpack loaded, one of the last things he would have done was tuck into his chest pocket a notepad, a pencil, and a hand lens that had been his father's.

At some point, Randy tore a page from a spiral notebook and wrote: "June 21: Ranger on patrol for 3–4 days. There is no radio inside the tent—I carry it with me. Please don't disturb my camp. This is all I have for the summer. I don't get resupplied. Thanks!"

He fastened the note to the canvas flap that served as his station's door, tightened the laces on his size 9 Merrell hiking boots, and pinned a National Park Service Ranger badge and name tag to his uniform gray button-down shirt. With an old ski pole for a hiking stick, he walked away from the station.

That afternoon, thunder rumbled across the mountains and raindrops pelted the gravelly soil surrounding his outpost, washing away his footprints and any clue as to the direction he had traveled.

IN SUMMERS PAST, Randy had anticipated boarding the parks' helicopter and flying into the backcountry with the giddy excitement of a child the night before Christmas. But this season had been different. The weather had grounded the parks' A-Star chopper for more than a week, which kept Randy and the other rangers on standby in what he called "purgatory."

Purgatory looked more like a UPS loading dock than it did an airbase at a national park. Dozens upon dozens of cardboard boxes were stacked haphazardly in waist-high piles waiting to be airlifted into the farthest reaches of the parks' backcountry. Each pile represented a ranger who had bought and boxed up three and a half months' worth of food and equipment that would last through the summer and into fall. Each box's weight was written in black marker adjacent to the ranger's name and the outpost that was its destination. Many of the veterans reused boxes year after year, so station names and weights had been crossed out numerous times, telling the story of their travels like tattered airline tags on the suitcases of frequent fliers.

Leaning against each pile of boxes was a backpack, maybe a duffel bag or two, and a crate of fresh produce—oranges, apples, a head of lettuce, a few avocados—the foodstuff that would be eaten first and missed the most on the rangers' tours of duty in the high country.

The men and women who loitered about wore hiking boots, running shoes, or the odd pair of Teva sandals, usually with socks. They were dressed in Patagonia fleece jackets, tie-dyed T-shirts, waterproof windbreakers, shorts—usually green, but sometimes khaki—worn over long underwear. The ensembles showed the duct-taped or sewn scars of prolonged use and were topped off by beanies, floppy hats, and perhaps one or two forest-green baseball caps with the embroidered NPS patch that betrayed their identities.

The average tourist might have pegged the group as a mingling of Whitney-bound mountaineers, dirt-bag climbers, and aging hippies. But make no mistake. These were America's finest backcountry rangers—Special Forces, if you will—disguised as an army of misfits. And most all of them were just fine with that description.

Not one of them wore the nostalgic cavalry-inspired hat so often associated with American park rangers. They weren't there to appear officious in head-to-toe gray-and-green uniforms; in fact, many of them were uncomfortable wearing a badge and carrying a gun. They weren't there to be wilderness cops, they were there to live and work in the wilderness, far from the roads their counterpart "frontcountry" rangers patrolled in jeeps and squad cars.

Some held master's degrees in forestry, geology, computer science, philosophy, or art history. They were teachers, photographers, writers, ski instructors, winter guides, documentary filmmakers, academics, pacifists, military veterans, and adventure seekers who, for whatever reason, were drawn to the wilderness.

In the backcountry, they were on call 24 hours a day as wilderness medics, law enforcement officers, search-and-rescue specialists, and wilderness hosts; interpreters who wore the hats of geologists, naturalists, botanists, wildlife observers, and historians. On good days they were "heroes" called upon to find a lost backpacker, warm a hypothermic hiker, chase away a bear, or save a life. On bad days they picked up trash, tore down illegal campfires, wrote citations, and were called "fucking assholes" simply for doing their job. On the worst days they recovered bodies.

The administrators in the park service often refer to them as "the backbone of the NPS." Still, they were hired and fired every season with zero job security. Their families had no medical benefits. No pension plans. And there was no room to complain because each one of them knew what they got into when they took the job. They paid for their own law enforcement training and emergency medical technician schooling. They were seasonal help. Temporary. In the 1930s, they were called "ninety-day wonders" who worked the crowded summer seasons.

Stereotypically, seasonal rangers were college students or recent grads taking some time off before starting "real" jobs. They would hang out in the woods for a few years and then move on, or start jumping through the hoops required to secure a permanent position in the National Park Service or Department of Interior. Sequoia and Kings Canyon, however, sucked in seasonal rangers like a vortex. More than half of the backcountry rangers who reported for duty in 1996 had been coming back each summer for more than a decade, many for two decades. Randy was the veteran, with almost three decades under his belt at these parks.

He was one of fourteen paid rangers budgeted to watch over an area of backcountry roughly the size of Rhode Island. Two of the rangers patrolled on horseback, the other twelve on foot.

These parks were two of the only national parks that still sent rangers into the wilds for entire seasons, and two of the few parks where these "temps" were more permanent than the "permanent" employees. Some of the park administrators called the SEKI (government-speak for Sequoia and Kings Canyon) backcountry crew "fanatics." Most of them were okay with that also. They were okay with just about anything as long as the weather would hurry the hell up and clear so the helicopters could transport their gear into the backcountry before their fruit began to rot.

As Randy milled about, waiting for the weather to clear, he sent mixed messages to his colleagues. By most accounts, he was "in a funk," "out of sorts," and conveyed little excitement for the season to come. The parks' senior science adviser, David Graber, considered Randy the parks' most enthusiastic and dedicated expert for "all things backcountry." He felt something was amiss when he saw Randy briefly at park headquarters at Ash Mountain. "I saw his big bushy beard coming from a mile away," says Graber, who had utilized Randy's expertise for virtually every backcountry-related scientific study he had supervised as the parks' ecologist for fifteen years.

They shook hands, and Graber—who had always counted on Randy for his passionate, curmudgeonly opinion on how the NPS wasn't doing enough to preserve his beloved backcountry—brought up the ongoing wildlife study they had been compiling for years and the current study on blister rust, a fungus that was spreading through the park, infecting and killing white pines. Randy didn't even entertain the topic. "Why bother?" he said with shrugged shoulders.

Graber at first assumed this blasé response had something to do with Randy's discontent with the park service, which was no secret. In the past, he'd conveyed that he felt backcountry rangers' duties weren't appreciated by the higher-ups in the park service—that they, like the backcountry itself, were being increasingly overlooked. "Out of sight, out of mind" was a popular cliché among the more veteran backcountry rangers, who said they put up with their second-class-citizen status in the National Park Service because of the excellent pay, a joke that would invoke a chuckle at any ranger gathering. It is an accepted truism that rangers are "paid in sunsets." After covering bills, gear, food, and the gas it takes to get their luxury automobiles—rusting Volkswagen vans, old Toyota trucks, and the like—to park headquarters, where they'd sit and leak oil till October, maybe a few dollars would trickle into a savings account. They certainly weren't there for the money.

In truth, there was one financial benefit backcountry rangers could count on. Randy, and all rangers with federal law enforcement commissions, was eligible for the Public Safety Officers' Benefits Program, enacted by Congress in 1976 to "offer peace of mind to men and women seeking careers in public safety and to make a strong statement about the value American society places on the contributions of those who serve their communities in potentially dangerous circumstances." In effect, the law offered a "one-time financial benefit paid to the eligible survivors of a public safety officer whose death is the direct and proximate result of a traumatic injury sustained in the line of duty." In 1976, the amount was $50,000; in 1988, that amount was increased to $100,000.

After twenty-eight years of summer service for the NPS, this was the only employment benefit Randy was eligible for. Of course, he would have to die first. So, here he was approaching his thirtieth year as a seasonal ranger at Sequoia and Kings Canyon and there was nothing about his uniform to distinguish him from a first-year rookie. There wasn't even a pin to commemorate the achievement: such medals were awarded only to permanent employees.

Graber, who had made it a point over the years to at least write letters of appreciation to the backcountry rangers for their invaluable contributions to his studies, had routinely told them that their job satisfaction "would have to come from within themselves—that they likely wouldn't get any from the NPS."

As Graber's conversation with Randy progressed, he interpreted the ranger's apathy and uncharacteristic lack of passion as depression. "His eyes were blank," says Graber, "but I knew how to push Randy's buttons—he'd lobbied for meadow closures his entire career. I never knew anybody who took a trampled patch of grass more personally than Randy. And wildflowers—he was a walking encyclopedia. You could always get him going about flowers, so I brought that up, along the lines of 'Nice and wet up high, good year for flowers.'

"His response was 'I don't find much pleasure in the flowers anymore.'"

That statement went beyond any contempt Randy held for the NPS. There was something else going on, but Graber didn't push the subject. "Randy wasn't the type to air his dirty laundry," says Graber, who patted Randy on the back when they parted ways. "I hope you have a good season, Randy," he said.

"You know, Dave," said Randy, "after all these years of being a ranger, I wonder if it's been worth it."

"That," says Graber, "chilled me to the core."

RICK SANGER WAS the tanned picture of a ranger in his prime—36 years old, 5-foot-11, with boyish good looks, dimples, and a quick smile. He'd quit a computer engineering job in 1992 and headed to the mountains for some healing perspective after the end of a stormy relationship. He was hired as a backcountry ranger on Mount San Jacinto in Southern California, where he stayed for three years before being hired in 1995 at Sequoia and Kings Canyon, parks he had been drawn to since his Boy Scout days.

This was Sanger's second season as a backcountry ranger in Kings Canyon. At dusk on July 23, 1996, he donned a headlamp, shouldered his backpack, and struck out into the cold outside his duty station at Rae Lakes. Randy Morgenson—stationed twenty miles north on the John Muir Trail—had been out of radio contact for three days, and it was Sanger's job to check on him. After a mile on the trail, Sanger's legs settled into a slow, steady, piston-like rhythm. With the cascading roar of Woods Creek on his right and towering granite peaks framing the starry-night sky, he couldn't believe he was getting paid to do this. God, he loved his job.

Sanger and Randy were a study in contrasts. Sanger was the young, gung-ho, clean-shaven newbie with a taste for adrenaline; Randy was the wise, weathered, and bearded sage of the high country who had pulled too many bodies out of the mountains to find any thrill in the prospect of a search-and-rescue operation. Sanger considered Randy a mentor for his uncompromising idealism in wilderness ethics. It had taken some time, however, to earn Randy's respect. The year before, the older ranger had studiously ignored him during training. Even when Sanger exhibited his expert mountaineering skills—self-arresting a fall with an ice ax on a snowy practice slope with the added difficulty of going headfirst while on his back—Randy had remained, at least outwardly, unimpressed.

The two were teamed up months later on a search-and-rescue operation and were forced to bivouac overnight in a steep gorge. Until dusk, Randy hadn't responded to Sanger with anything more than yes or no as they searched for a missing backpacker. The silence was undoubt-edly enjoyable for Randy, but offensive to Sanger, who interpreted it as rudeness. As darkness settled, Sanger gathered some wood for a small fire. After an entire day together, Randy uttered his first complete sentence: "You'd do well to learn a little respect."

Sanger was at once offended, confused, and angry. He had been trying to engage in conversation all day, and this was Randy's recip-rocation?

"And in what way have I not been showing you respect?" asked Sanger. "I've been wanting to work with you all day, to learn from you. I don't think you realize the regard I have for you and your experience in these mountains."

"No, Rick," said Randy. "I'm referring to the fire."

Randy moved his tiny backpacker stove closer to where Sanger sat, squatted beside him, and explained why Sanger should not build a fire—even though the wood he'd intended to burn was already dead; even though they were at a legal elevation for campfires; even though the blackened residue from the fire on the rocks and sand would be washed clean the next rain cycle. What gave human beings—not to mention rangers—the right to alter the natural processes at work here?

Sanger respectfully scattered the wood he had gathered, and in doing so earned the regard he was seeking and kindled a friendship. A mentorship in wilderness ethics was born. Over the course of the night, Randy opened up and offered Sanger a rare glimpse inside the backcountry rangers' most notorious recluse.

On subsequent contacts, the bond had continued to grow. Sanger knew Randy was working his way through some issues—unfinished business with his father as well as a marriage that was on the rocks—but he also knew that the backcountry had amazing healing properties. Randy had even told the younger ranger, "There's nothing a season in the backcountry can't cure."

Now, as Sanger hiked through the night toward Randy's station, he looked forward to the ritual of boiling a kettle of water and catching up over cups of tea. When he had delivered a new radio to Randy at the White Fork trail-crew camp a couple of weeks earlier, Randy had seemed excited about the future and hadn't exhibited any signs of the depression reported by other rangers.

At the White Fork camp, Randy had been reading *Blue Highways* by William Least Heat-Moon, an account of the author's 11,000-mile road trip instigated by some setbacks in his life, including marital problems. The introduction to *Blue Highways* reads:

> On the old highway maps of America, the main routes were red and the back roads blue. Now even the colors are changing. But in those brevities just before dawn and a little after dusk—times neither day nor night—the old roads return to the sky some of its color. Then, in truth, they carry a mysterious cast of blue, and it's that time when the pull of the blue highway is strongest, when the open road is a beckoning, a strangeness, a place where a man can lose himself.

Sanger was curious about whether Randy had maintained the level of optimism he'd expressed in the frontcountry when he'd half-seriously, half-jokingly told Sanger that he had been thinking about trying something new: "Maybe I'll try my hand as a river guide or a racecar driver." Sanger and another backcountry ranger subsequently dubbed him "Maserati Morgenson." But Sanger couldn't imagine Randy as anything but a backcountry ranger—and, selfishly perhaps, wanted him to stick around for a while.

True to his private nature, Randy hadn't shared with Sanger, or any of his fellow rangers, the unwanted burden he had brought upon himself: the divorce papers his wife sent with him into the backcountry. He was a signature away from ending his marriage of twenty years.

Perhaps that was what Randy was thinking about when he'd told Sanger at the White Fork, "Few men my age have the freedom I've been afforded," following with "The sky's the limit." But he never brought up the divorce papers. "He seemed," says Sanger, "to be exploring the options for his future—and using me as a sounding board."

When Heat-Moon got the idea to skip town, he wrote: "A man who couldn't make things go right could at least go. . . . He could quit trying to get out of the way of life. Chuck routine. Live the real jeopardy of circumstance. It was a question of dignity." It certainly sounded romantic on paper, but it hadn't been easy for Heat-Moon. He wrote of lying awake at night, tossing, turning, and "doubting the madness of just walking out on things, doubting the whole plan that would begin at daybreak."

Was it purely coincidental that Randy had been reading this book, and seemingly dropping hints about starting a new life, just two weeks before he disappeared?

ON THE MORNING OF JULY 24, Sanger was head down and pounding the switchbacks up 12,100-foot Pinchot Pass—hoofing it "big time" to make the summit by 11:30 for the morning roundup, when park headquarters checked in via radio on all the backcountry rangers. The Pinchot Pass ridgeline was the border between his patrol area to the south and Randy's to the north, but this morning its lofty perch would serve as a craggy granite radio tower from which Sanger would send a signal—unimpaired—to the Bench Lake station 4 miles north and 2,000 vertical feet below in the mountain-rimmed Marjorie Lake Basin. Randy, he reasoned, might be having problems reaching park headquarters far to the southwest, but would nonetheless be monitoring during roundup. From the pass, Sanger's transmission would be loud and clear for anybody in the area.

Barely making it in time, Sanger transmitted, using Randy's radio call number, 114.

"One-one-four, this is 115 . . . 114, this is 115. . . . Hey, Randy, you out there?"

He persisted, trying all the channels used by the parks. When he was certain nobody was there, he contacted the parks' dispatcher, who confirmed that Randy was still unaccounted for.

The last time Randy checked in had been four days earlier, on Saturday, July 20, from Mather Pass, six and a half miles north of his station on the John Muir Trail. Eric Morey, the Grant Grove subdistrict ranger, had performed morning roundup that day and later recalled that Randy's "radio communications were poor" and that he "might have said something about his radio batteries working poorly."

But why, considering the parks' backcountry-ranger safety policy, had it taken four days to get a ranger into Randy's patrol area? In this case it would prove to be a breakdown in communications of a different kind. The protocol clearly stated:

> Due to the remote locations that backcountry rangers are assigned . . . in order to provide for their safety . . . radio communication will be made daily . . . at 1130 hours. If communications cannot be made . . . it will be noted in the status book. If communications still have not been made within the next 24-hour period . . . the employee's supervisor will be notified and further efforts to locate that ranger will be initiated.

But what if the employee's supervisor—in this case Sierra Crest Subdistrict Ranger Cindy Purcell—was on vacation? There was no written policy for that scenario. And so "N/C" (no contact) was written next to Randy's name on the backcountry radio log for three days in a row. Purcell's supervisor, District Ranger Randy Coffman (the man who had written the protocol), was finally informed of the situation by the district secretary, Chris Pearson. Pearson, who sporadically performed morning roundup, noticed that Randy had not been in contact for three days. Since Purcell was out of the park, Pearson felt "somebody should know."

Coffman acted immediately and contacted Sanger late in the afternoon of July 23, during a prearranged time when rangers were expected to monitor their radios. It was then that Sanger's patrol, officially noted as a "welfare check" to Bench Lake, was initiated.

None of those details mattered to Sanger. As far as he was concerned, it was just another beautiful day to patrol in the high country. Checking on another ranger, Randy in particular, was the icing on the cake.

The likelihood that anything had gone wrong was practically nil in his mind. And besides, Coffman, the parks' preeminent search-and-rescue expert, couldn't have been overly concerned; otherwise, he wouldn't have sent Sanger nearly 20 trail miles on foot, knowing that he wouldn't arrive at Randy's duty station until the following day. The parks' helicopter could have transported a ranger to Bench Lake in less than 30 minutes.

"I was no more concerned about [Randy] than I was when my ex-girlfriend's cat stayed out all night," wrote Sanger about his mindset that day. "Not in the sense that I don't give a hoot about cats, but that I believe implicitly that cats can take care of themselves."

Further illustrating Sanger's lack of concern, he took advantage of the altitude to call his father on his modified ham radio, which was also a radio telephone, and wish him a happy birthday before he descended from the pass.

But before taking the first step into Randy's patrol area, Sanger's recent law enforcement training switched on. Despite his optimism that everything was okay, something heinous could have happened. If some threatening, potentially violent individual was in the area, Sanger reasoned it best not to approach the station in uniform. He changed into plain clothes and headed toward Randy's station, hopeful that his precautions wouldn't be justified.

As the trail passed the deep blue waters of Marjorie Lake, Sanger's strides lengthened. Except for the cheerful banter of Clark's nutcrackers darting back and forth from the tops of altitude-stunted lodgepole pines, everything was quiet. It was a spectacular day in the high country.

The trail leveled out in an alpine meadow and paralleled a creek for a couple hundred yards before intersecting with the Taboose Pass Trail, which was a rock-hop over the creek. A few yards later, a metal sign planted in the gravelly soil read RANGER STATION. Along a barely perceptible footpath through some scattered lodgepoles, Sanger approached the tent casually.

"Hello, anybody home?" he called out from a distance.

Silence.

At the station's door he read the note Randy had left four days earlier and did the math. If all was well, Randy should be walking into camp at any time. He relayed this to Coffman and suggested waiting until evening before starting a search. He was certain Randy would show up; in fact, he felt uncomfortable entering Randy's private living quarters. But he did enter, per Coffman, to look for any clues—perhaps a patrol itinerary—that might shed light on the unaccounted-for ranger's whereabouts.

Sanger reported back to Coffman that everything was in order inside the tent, and that no itinerary was present or mentioned in the station logbook.

An alarm went off in Coffman's brain. He consulted briefly with Dave Ashe, the acting Sierra Crest subdistrict ranger and Randy's supervisor the season before. Ashe knew Randy attracted bad radios like a magnet.

"I didn't want to rush into anything," says Ashe. "I just figured he'd show up on a trail if we started a search right then. I thought we should let the full four days play out first."

Coffman ignored Ashe's and Sanger's instincts to wait and checked the availability of the parks' helicopter, known by its radio call number, 552. Within minutes, he had coordinated a flight plan for himself and a handful of rangers to rendezvous with Sanger at the Bench Lake ranger station.

While Coffman prepared his gear and made his way to the heli base, the park dispatcher attempted to contact three backcountry rangers: George Durkee, Lo Lyness, and Sandy Graban. The choices weren't random; Coffman knew that all of them were longtime friends of Randy and each was familiar with the Bench Lake patrol area. A handful of other rangers were subsequently alerted to the situation and placed on standby.

The radio communication was concise: Pack a backpack for three days and head to the nearest landing zone—a search-and-rescue operation was in progress for 114.

BACKCOUNTRY RANGER GEORGE DURKEE was removing a fallen tree from the trail switchbacks high above his LeConte Canyon ranger station when he got the call. The 6-foot-2 ranger with a distance runner's physique had become known as "The Commander" both for the high-water jumpsuit he wore during training and for his ability to bite his tongue and be the smiling, red-bearded diplomatic voice of the back-country rangers. He describes himself as an "aging hippie who moved with the speed and grace of a creaky cheetah."

Genetically incapable of not inserting humor into almost any situation, Durkee had recently made himself "Sequoia Kings Canyon, National Park Service" business cards. The cards prominently displayed a flashy gold NPS badge with his name and the words "Park Ranger" centered above the slogan "Manly deeds, manfully done."

Despite his class clown tendencies, Durkee was a hardened veteran of the ranger ranks. In the early 1970s, he'd been known to "stalk the SAR cache" in Yosemite, where his career with the NPS began. The SAR (rhymes with car) cache was the quick-access search-and-rescue storage facility for emergency medical supplies such as backboards, ropes, litters . . . and body bags. Between 1972 and 1977, Durkee assisted in the recovery of more than twenty-five bodies. It was during this SAR junkie phase of Durkee's life that he'd met Randy, ten years older and at the time a Nordic ski ranger stationed out of Badger Pass, Yosemite's ski area. Their friendship was born of a mutual love of wilderness and a sardonic sense of humor.

Now 44, Durkee hadn't lost his taste for adrenaline, but it had begun to ebb and flow, depending on the level of the catastrophe. Same with his friendship with Randy, which only recently had become "strained."

At the time of the radio call, Durkee was forty minutes from his station. He dropped what he was doing and hoofed it to his cabin in twenty minutes, stuffed three days' worth of food into his backpack, kissed his wife—volunteer ranger Paige Meier—and was pacing at the designated helicopter-landing zone in less than an hour. He was concerned for his friend, having been privy to some of the personal issues Randy was dealing with. As he waited, three particular memories repeated themselves.

First was the time he and Randy almost simultaneously met their ends at the blades of a military helicopter's rotor while rescuing two hikers on Mount Darwin on August 20, 1994. One of the climbers was trapped on a ledge and the other was severely injured after falling 140 feet down a steep snowfield. It was precarious, you-slip-you-die terrain, with few helicopter-landing zones and lots of wind. A gust spun the tail of the helicopter, causing the main rotor to lurch dangerously close to some protruding granite just above where they were huddled around a litter on an indentation of Darwin's northern slope—trying to hoist the injured climber into the chopper and to a hospital. Just the thought made Durkee duck.

The rescue was a success, but as Durkee had moved down a rocky couloir, he knocked loose a rock the size of a softball. He yelled "Rock!" an instant before it hit Randy squarely on the head, knocking him senseless. If it hadn't been for the helmet, Randy probably would have died.

They earned an award for small-unit valor for that rescue. It had been the second of only two awards for exceptional service Randy received from the Park Service during his entire career. In his personal report of the rescue, Randy never mentioned the rock that Durkee had knocked loose. He hadn't wanted the incident to reflect poorly on his friend.

After they got the climbers out of the mountains, the park helicopter had picked them up at the base of Darwin and flew them to McClure Meadow. The two found a comfortable flat spot and lay on their backs watching the clouds, still feeling the adrenaline coursing through their veins. Durkee commented on what a great day it was to be alive. Randy's response had been "Oh, I don't know." He sat up and scanned the meadow and the mountains that rose up from Evolution Basin—spectacular peaks named after Darwin, Huxley, and other evolutionary thinkers. And then he said, matter-of-factly, "The least I owe these mountains is a body." By itself, that remark was more maudlin than suicidal, but when a man disappears in those same mountains to which he has said he owes a body, a friend starts adding up the clues.

The second memory was an argument Durkee had had with Randy during training the year before, in June of 1995. A low-key conversation had escalated and Durkee released a boatload of pent-up resentment that had been simmering for more than a year about an extramarital affair.

"Whether it was a midlife crisis, filling a void, or just a side of Randy I didn't know existed, he was hurting his wife, who was also my friend," says Durkee. "Not only had he put me in an extremely difficult position, he was also losing my respect, so I told him so."

Randy lashed out verbally and told Durkee he was being judgmental. Durkee countered by telling him he was only judging the pain he had been causing Judi. "Don't you think I know I'm causing Judi pain?!" Randy erupted. "I was this close"—thumb and forefinger a centimeter apart—"to not coming back this season!"

Then Randy sat down and started to cry, his face in his hands. But he quickly composed himself and admitted that after Judi had found out about the affair, he'd started thinking about suicide. "Not seriously," Randy assured Durkee, "but I've been having those kinds of thoughts."

The third memory was from July 20, 1996—the night before Randy went "on patrol"—when he had radioed Durkee and Meier, asking some mundane questions that Durkee interpreted as "Randy just wanting somebody to talk to." The short radio conversation had ended when Randy said abruptly, "I won't be bothering you two anymore." Durkee and Meier looked at each other with the same "He wasn't bothering us" expression and shrugged it off.

Now, with his friend missing somewhere in the backcountry, Randy's words were deeply troubling. Durkee couldn't wait to get to Bench Lake, not only to start the search but also to see if Randy had taken along the Smith and Wesson .357 Magnum he'd been issued for the season.

The decidedly heavy two pounds of steel plus ammo was a required part of the uniform. But Durkee knew that Randy always left it locked up at his station while on off-trail patrols. He despised the gun for what it did to the once approachable park ranger uniform, and had conveyed serious doubts about being able to pull the trigger against another human being, even in self-defense. If the gun wasn't at the station, Durkee feared that Randy might have had plans to use it on himself.

LO LYNESS—the Charlotte Lake ranger—had, for the past few years, been closest to Randy, though they hadn't intended the intimacy of their relationship to be public knowledge. But like any workplace, a national park isn't devoid of gossip, so most of the backcountry rangers knew of their affair. It had ended recently, but she still held deep feelings for the man.

Perhaps it was the bond that they'd formed in that relatively short but intense relationship that had alerted her to some unexplained distress two days before the search-and-rescue operation for Randy was initiated. She was off-trail near Upper Sphinx Creek when she'd heard on her radio that Randy had not checked in for a couple of days. Her intuition, even then, told her that "something was truly wrong."

For the 5-foot-10, blond, fair-skinned Lyness, there was magic in these mountains. After a couple of weeks in the high and lonely, all the backcountry rangers experienced a slowing down. Randy called it "decompression," a transition from the fast pace and crowds of civilization. Once in wilderness, a Zen-like calm heightened their senses exponentially with each passing day. Even skeptical rangers admit that an unmistakable zone comes with time and solitude. Randy had likened the quieting sensation to religion—"a theology not found elsewhere," he wrote in his logbook while stationed at Charlotte Lake in 1966. He had struggled then to explain these "Sierra moments . . . only experienced when still . . . and surrounded by and conscious of the country."

In 1973, he wrote more extensively on these feelings while stationed at McClure Meadow, his most cherished meadow in the park. There, he sensed he was "close to something very great and very large, something containing me and all this around me, something I only dimly perceive, and understand not at all."

"Perhaps," he pondered, "if I am here, aware, and perceptive long enough I will."

Lyness was in that heightened state of awareness, and though she knew it wasn't unusual for a ranger to be out of contact for days at a time, she couldn't deny being unusually "anxious and disturbed."

On July 24, about the same time Sanger was approaching Pinchot Pass, Lyness left her cabin on patrol. By morning roundup, a few hours later, her radio had died. She continued on her patrol to Vidette Meadow, checking in on campsites and performing the rangers' most menial, backbreaking jobs of cleaning and deconstructing illegal or oversized fire sites—a sure way to physically work the worry out of her. In the afternoon, tired from moving rock and covered with soot, she made her way home. While on the trail, she saw the parks' chief ranger, Debbie Bird, on horseback. Lyness, anxious to know if Randy had checked in that day, asked Bird the status, but Bird, who had been off-duty in the backcountry with her family, wasn't even aware that Randy was incommunicado. As they spoke, gray clouds and a slight sprinkle hastened their conversation, but even in the span of a few minutes, Bird noted Lyness's "obvious concern for Randy." She was on her way out of the backcountry, so she gave Lyness her radio, then watched the willowy ranger stride toward the tree line and the switchbacks that led to her cabin high above the wooded slopes surrounding the meadow.

Though she didn't mention anything to Lyness, Bird felt a sense of foreboding in the gathering storm. "It might have been the turn in the weather—thunder was grumbling in the distance—but something was in the air that day," she says.

Fifteen minutes after she parted ways with the chief ranger, Lyness's newly acquired radio died. Soon thereafter, she heard a helicopter, which for a ranger usually meant trouble.

On the switchbacks that climbed nearly a thousand vertical feet from the meadow to the trail junction to Charlotte Lake, Lyness saw the park helicopter in a wide circling pattern above her station, obviously looking for her. Standard protocol was to use a direct channel to contact the park helo if it passed nearby. She cursed the radio and ran up the steep switchbacks, almost making the summit as 552 flew directly overhead and away. Already exhausted, she continued past her cabin to the trail-crew camp at the far end of the lake and used the camp supervisor's radio to call the helicopter back. Giving her just enough time to organize her backpack, the helicopter touched down and whisked Lyness north toward the Bench Lake station.

IN A CURSORY, "hasty" air search, the rangers flew into Randy's patrol area from their outposts in the south, west, and north. Each of the rangers used channel 1, the park's direct radio line, to transmit blind messages to Randy while scanning the alternating granite, meadow, and wooded terrain below. If he was conscious, but injured and unable to get to a spot where his radio could hit a repeater, he'd only have to turn on his radio to make contact with the overhead helicopter.

"One-one-four, 114—Randy, we are starting to search for you. Head back to your station or get out in the open where we can see you."

"One-one-four, 114—Randy, we are starting to search for you." Different versions of essentially the same message were repeated again and again on all of their overflights.

There was no response.

Once on the ground at the Bench Lake station, the team of rangers felt an immediate hole in their ranks. Durkee describes their gathering as "the unusual suspects." The "usual suspects," he says, "would have included Randy."

There were no pleasantries, but hands were shaken and hugs exchanged before they got down to business.

Fifty-year-old Sandy Graban, the park's most senior female back-country ranger, with nineteen seasons under her belt, stood comfortably at one end of the picnic table where Randy generally ate his meals. She would admit later that she thought they were "jumping the gun."

"Randy wasn't missing," she explains, "he was overdue—and had been numerous times before."

Tall and powerfully built after years of carrying a heavy backpack, Graban had attended a ranger law enforcement academy with a bunch of twenty-somethings when she was 40. Despite the generation gap, she had graduated at the top of her class in physical fitness.

Colleagues describe Graban's thoughtful and "spiritual" persona as having the ability to slow everything down. But on the trail, she would hit warp speed and leave most rangers in her dust. Sanger, a keen observer and never without a notepad, had noticed in his short tenure at the park that Graban generally sat and listened at the edge of conversations, but "her capabilities and experience were evident when the group eventually deferred to her judgment." Despite Graban's connection with the mountains and longtime friendship with Randy, nothing had kinked her senses. No "bad vibes" were reverberating from the granite. The only thing she had noted during training was how "Randy's mood had seemed 'heavy.' " Otherwise she, like Sanger, felt that "Randy could take care of himself."

All present half expected their friend to come walking up at any moment, white teeth smiling through the familiar bushy salt-and-pepper mountain-man beard, with a remark like "Who's the party for?" He would toss a broken radio on the table with a snarl of contempt and grunt while unshouldering a pack that was heavier than it was when he'd left, now filled to capacity with "backpacker detritus," his term for tinfoil, candy wrappers, beer bottles, and the like. "Cleaning up after grown men," he'd been known to say. "A never-ending battle."

That fantasy evaporated with each passing minute.

Once Randy Coffman sat down, all the rangers converged around the picnic table and gave him their full attention.

"In any gathering, Coffman was the alpha male," says Durkee. "If there was any doubt for those visiting his office in the frontcountry, a photo of a huge grizzly bear behind his desk was an apt reminder." A skilled mountaineer with high-altitude ascents in the Andes, Alaska, and Africa, the muscular 5-foot-9 district ranger had been on the summit team for the

1994 American-Norwegian International Expedition to the thirteenth-highest mountain in the world, 26,360-foot Gasherbrum II in Pakistan.

Coffman taught courses in the art of search and rescue and was considered the resident SAR expert at Sequoia and Kings Canyon. In the tangled red-tape bureaucracy of the frontcountry, Coffman was a technically perfect ranger building a résumé that would eventually land him a high-level NPS position in a Washington, D.C., office. But despite extremely capable field skills, Coffman was described by his colleagues as "arrogant, autocratic, and oftentimes difficult to work with."

As one ranger says, "You couldn't be in a room with Coffman for more than five minutes before he pissed everybody off."

But in the backcountry, Coffman was a different person, and in a crisis he both thrived on the intensity and relaxed, seeking out and welcoming input as he wrapped his mind around the situation. In a search-and-rescue operation, those same people who had berated his interpersonal-management skills couldn't think of a more talented or qualified person in the park to lead the effort.

In an SAR, it was unanimous: Coffman shined. And it showed as he calmly and confidently spearheaded what would become one of the most challenging SARs in his career, made more difficult because, like the others, he considered Randy Morgenson a friend.

Shortly, Coffman would learn that Randy was at a crossroads in his life. As such, the four-way intersection of dotted red lines he'd been staring at on the map represented far more than just trails leading away from the Bench Lake ranger station.

East of the creek, the Taboose Pass Trail continued northeast—the quickest, most direct route out of the mountains: 23 miles to Highway 395. West of the creek, the Taboose Pass Trail became the Bench Lake Trail, which dead-ended 2.4 miles later at the west end of Bench Lake, where a myriad of Randy's preferred cross-country routes led to some of his most cherished hideaway mountain basins. South, the crowded John Muir Trail traveled 59 miles and terminated on the summit of Mount Whitney. One hundred fifty-two miles north was Randy's childhood home, Yosemite Valley, where this story really began.

CHAPTER TWO

THE GRANITE WOMB

[We] moved to Yosemite Valley and settled into the
house assigned to us, which faced Half Dome and the
rising sun. We felt that the sun had risen permanently
in our lives.

—*Esther Morgenson, 1944*

Earth laughs in flowers.

—*Ralph Waldo Emerson, "Hamatreya"*

IN 1950, THERE WAS no real trail penetrating the stark high-alpine
landscape surrounding Mount Dana in Yosemite National Park. To the
casual eye it looked as if no living thing existed at these heights, except
for the two specks making slow progress, climbing among the loose
talus and boulders on Dana's western slope.

One of the climbers, 8-year-old Randy Morgenson, was within a
few hundred yards of the summit of his first 13,000-foot peak. The
lack of oxygen slowed the boy's progress to a snail's pace, frustrating
him slightly because he couldn't just run up this mountain the way
he did the trails down in Yosemite Valley, where he lived. His father,
Dana Morgenson, a few steps behind, explained the effects of altitude
but focused cheerfully on the benefits of a slow pace, admonishing his

wiry son to take advantage of his breathlessness to enjoy the view and notice the deceptive living garden in which they'd paused—granite slabs covered by red, orange, and gold lichens. These were the first striking colors they'd seen since rising above the lodgepole pines and meadows they'd passed through hours earlier.

After a short break, the tired young mountaineer was rejuvenated by his father's promise of a rare treasure at the rocky summit.

With less than 500 feet to go, little Randy investigated rock overhangs and tiny crevices as he crept slowly upward. In a shady alcove he discovered a tiny patch of golden flowers growing in a sandy flat. He hollered with delight at the discovery and called his father to identify the find. Dana bent down beside his son and focused his camera on the first reward of the day's hike: *Hulsea algida*, otherwise known as alpine gold. Next, Dana took an ever-present hand lens from his shirt pocket and revealed to his son the amazing, magnified world of nature.

At the rocky summit, Randy lay on his belly to breathe in the floral scent of Mount Dana's most remarkable treasure: the pale blue *Polemonium eximium.*

Dana told his son that the common name of this flower is sky pilot, so named because it is found only on or near the tops of the highest peaks. "The name," said Dana, "means 'one who leads others to heaven.' " With wide-eyed excitement, Randy reached to pluck a tiny bouquet for his mother, but his father stopped him, explaining how the delicate flowers had fought long and hard to survive in such a harsh environment. He then posed the question, "Wouldn't it be nice to leave these alone?" He explained in terms an 8-year-old might grasp: if climbers before them had picked these flowers, they wouldn't now be enjoying their beauty.

Upon arriving back at their cabin in Yosemite Valley, Randy ran into the kitchen, where his mother, Esther, was preparing dinner. "Mother, Mother!" he exclaimed. "I found pandemonium!" He didn't understand why his mother and father found this so funny, even after they corrected the flower's name. Randy's hunt for "pandemonium" became an oft-repeated Morgenson family story.

◑

TO UNDERSTAND WHO Randy Morgenson was and, more important, *why* he became who he was, one need look no further than his father.

Dana Morgenson was born in the Midwest, and when he was a child, his family moved to the town of Escalon, in the Great Central Valley of California. Shortly thereafter, Dana's gaze was set toward the high and mysterious mountains of the Sierra Nevada. Even after he graduated from Stanford University in 1929 with a degree in English, the mountains' magnetic pull had not subsided.

The Great Depression struck, and Dana was happy to find employment at the same bank in Escalon where his father had worked. It was in 1930 that he earned his first paid vacation. With a friend, he decided to finally explore the Sierra, camping in Yosemite National Park's Tuolumne Meadows. He brought with him a fishing pole because, he thought, "That is what one does in the mountains." As an afterthought, he also carried along a Brownie camera. Within a day, the fishing equipment was cast aside in favor of photographically documenting the experience. He and his friend fished, hiked, and climbed, coincidentally, Mount Dana, on which Dana Morgenson would later do research and discover that it had been named during the 1863 California Geological Survey for James Dwight Dana, the foremost geologist of his time. To Dana Morgenson, the peak was simply the top of the world.

Back in Escalon, Dana was disappointed by the quality of the images he'd taken with the rudimentary box camera. He purchased photography instruction manuals and longed for the better equipment that the lean years of the Depression wouldn't allow.

While Dana was working at the bank, a girl named Esther Edwards, whom he had known in childhood, caught his attention. She'd moved away during grammar school, but had recently returned to study art at a nearby college. They were smitten with each other.

On September 9, 1933, the two 24-year-olds were married in a simple garden wedding at a friend's house. "The most beautiful day of my life," wrote Dana in his diary. "Esther was indescribably lovely and I was supremely happy!"

After a 12-day camping honeymoon up the California coast, they returned to Escalon, where Dana worked long hours at the bank and Esther earned her degree. Vacations to the mountains and deserts were marked well in advance on the house calendar, and Dana stayed in shape by running up and down the stairs to the bank's upper level during his lunch break. They took weekend camping trips to explore the Sierra whenever Dana could manage a Saturday off from the bank—which wasn't often enough.

He was an avid journal keeper and reader. A favorite on his bookshelf was a second-edition *Guide to the Yosemite Valley,* published by "authority of the legislature" in 1870. The book, whose contents were the work of the Geological Survey of California, was illustrated with detailed maps and woodcuts, all of which fueled Dana's burning desire to make a living and raise a family in a wilderness setting. Dana asked his wife on numerous occasions, "Wouldn't it be nice to live where you could walk in the woods on a Sunday afternoon?" It became Esther's dream also.

But in the 1930s, nobody left a good job to chase such romantic notions.

Their first son, Lawrence (Larry) Dana Morgenson, was born on April 12, 1938. James Randall (Randy) Morgenson arrived on May 21, 1942, not long after the United States declared war on Japan and entered World War II. Before Randy was 1 year old, he had been baptized into the world of camping in Yosemite by being bathed in a campfire-warmed bucket of water dipped from the Merced River.

In the early 1940s, Yosemite was a snapshot of American life, with an area of the park set aside for "victory gardens" and uniformed Army and Navy soldiers and sailors billeted in semipermanent camps around the valley, including the Ahwahnee Hotel, which was also used by convalescing soldiers. In 1944, D-day signaled the beginning of the end of the war, and with the anticipated end of gas rationing, the National Park Service and its concessionaires prepared for a surge in visitors. As a result, Dana was offered a job in Yosemite National Park as the office manager for its concessionaire, the Yosemite Park and Curry

Company. The following day, Dana gave notice at the bank.

In August, the family moved into House 102 on Tecoya Row, Yosemite Valley—the Curry Company's employee housing area. The small but comfortable clapboard half-a-duplex was, for Dana and Esther, a romantic wilderness cabin. After more than a decade of uninspiring desk work at two banks and years of discreet yet persistent job hunting, Dana had realized his dream to live and work in the mountains. It was still a desk job, but that was of little consequence. House 102 came complete with living and dining room windows overlooking the waving tall grasses of the Ahwahnee Meadow. Towering over a wall of trees across the meadow stood the awe-inspiring granite monoliths of the Royal Archers, Washington Column, and—dominating the horizon—the world-famous Half Dome. Their back door opened to yet another dizzying wall of granite, which rose above the evergreen forest that surrounded the home on three sides. It was a wilderness utopia enhanced by the sound of the rushing Merced and the nostalgia-inducing smell of piney campfires each evening. Friends and relatives who came to visit commented that the Morgensons were living in a postcard.

Once settled, Dana spent every spare moment exploring the mountains and "learning their secrets," he'd often write in his journal. His passion became the wildflowers. He read avidly and befriended local naturalist Mary Tresidder and the renowned naturalist Dr. Carl Sharsmith, who sensed in the man from the accounting department an unexpected kindred spirit for the park's wild places and shared with him their knowledge of the park's secret gardens.

Dana garnered his own reputation as the valley's authority on wildflowers, which in time would be his ticket out of the Curry Company's accounting office. Beginning each spring, he was approached to identify flowers or point park employees or tourists down the right mountain trail which he had, over the course of years, walked, documenting all the species with both his journal and his camera. Shadowing Dana on many of these outings were his two boys, both of whom were casually taught the scientific names of wildflowers and trees on one walk, trail names and peak heights on the next. Always on these adventures, they were fed a seemingly endless diet of quotes from John Muir, Albert Einstein, Henry David Thoreau, Walt Whitman, and Ogden Nash—a few of the authors whose books lined the walls of the Morgenson home, which came to be known as one of the valley's more extensive private libraries.

One of Dana's favorite Whitman quotes was "To me, every hour of the day and night is an unspeakable perfect miracle." It was this sort of appreciation of their surroundings and life that instilled a sense of awe in Dana's two sons, who especially liked walks along the rushing torrent of the Merced River. "Thousands of joyous streams are born in the snowy range," Dana quoted Muir, "but not a poet among them all can sing like Merced."

The Morgenson brothers learned from their father that church and wilderness were one and the same. Though they regularly attended Sunday services in the valley, Dana wouldn't think twice about replacing a pew with a chunk of granite on a Sunday morning hike to, for example, the wet and boggy Summit Meadow in search of the "ghostly" white Sierra rein orchid or, as he would record in his notes, "*Habenaria dilatata* of the *leucostachys* variety."

Dana would talk to the animals of the forest as though they were neighbors, saying "Good morning, Mrs. Squirrel, how are the kids?" when passing a trailside burrow known to house a litter of pups. Equally respectful to the two-legged fauna inhabiting the park, he would tip his hat to park rangers, which no doubt made an impact on Randy, whose favorite poem as a youth was "Ranger's Delight." The humorous amateur poem, found in a book on the Morgensons' bookshelf entitled *Oh Ranger,* by Horace M. Albright and Frank J. Taylor (1928), was bookmarked by Dana with a slip of paper on which he'd scribbled "Randy's favorite."

The season's over and they come down
From the ranger stations to the nearest town
Wild and woolly and tired and lame

From playing the "next to Nature" game.
These are the men the nation must pay
For "doing nothing," the town folks say.
But facts are different. I'm here to tell
That some of their trails run right through—well,
Woods and mountains and deserts and brush.
They are always going and always rush.
They camp at some mountain meadow at night,
And dine on a can of "Ranger's Delight,"
They build cabins and fences and telephone lines,
Head off the homesteaders and keep out the mines.
There's a telephone call, there's a fire to fight;
The rangers are there both day and night.
Oh, the ranger's life is full of joys,
And they're all good, jolly, care-free boys,
And in wealth they are sure to roll and reek,
For a ranger can live on one meal a week.

The poem, reportedly written by someone known only as "Canned Tomatoes," was said to have been found in a ranger cabin in El Dorado National Forest around 1928. Randy's taste for literature matured with his years, and he quickly graduated from "Canned Tomatoes" to many of the same authors his father quoted with ease. Soon enough, Randy, too, was quoting Thoreau and Muir from memory, and family and close friends nodded their heads knowingly. It was obvious that the cone hadn't fallen far from the pine tree.

THE YEAR-ROUND RESIDENTS of Yosemite often referred to their valley as a "granite womb." Shielded from the problems of city life, they didn't lock doors. Keys were left in the ignition or atop the sun visor in the car, and children weren't limited by backyard fences. One of Randy's childhood friends was Randy Rust, the son of the postmaster. Rust remembers when kids walked around with bows and arrows, BB guns, and fishing poles. "We never shot anything but cans," says Rust.

"The big difference back then was that when we saw a ranger, he'd stop and shake our hands and check out our weapons, talk to us like we were real mountain men—and then be on his way with a tip of his hat. Today, if a ranger saw a kid walking around the valley with a BB gun, that gun would be confiscated in a second."

As teens, they'd "float down the Merced in old inner tubes, and fish," says Rust. "Nobody had television in the valley till we were in high school and radio reception was horrible. Sometimes we'd all gather at different houses—the Morgensons were one of the families with a phonograph—and we'd listen to records. Sometimes Mrs. Morgenson would be painting in the front yard, and sometimes she'd make lemonade for us with a pitcher and glasses, served on a tray. The Morgensons were very proper."

Randy walked or rode his bicycle a quarter mile to the two-room schoolhouse on meandering pathways where he would often get "lost" after school, barely making it to the dinner table in time for the carving of a ham or meat loaf. That is, unless some guest was joining the family for happy hour before dinner, during which the adults would enjoy a cocktail or two—in front of the fire in winter or loitering in the front yard watching the shadows creep across Half Dome in the spring and summer. Randy, an eager listener, was rarely late when guests like Ansel and Virginia Adams, or some other distinguished Yosemite visitor whom his parents had befriended, was expected. Often, Randy was requested to choose the evening's music. He'd gladly set to the phonograph some classical record fitting of the weather or mood. While many of his teen peers were busy wearing out Elvis Presley's new single, "Don't Be Cruel," Randy remained drawn to classics such as Artur Rubinstein's rendition of Grieg's Piano Concerto in A minor.

With or without a VIP guest, dinner was always a sit-down affair at the Morgensons'. Esther mimicked her English mother's regimen of a perfectly set table. The meat was carved by the man of the house, milk was served in a pitcher, and hats and elbows weren't tolerated.

Entertainment after the evening meal was generally focused on conversation, either around the fire or at one of the venues in the valley where visiting scholars, authors, artists, and photographers frequently gave lectures and slide shows and presented documentary films. In later years Larry would veer off to a high school party while Randy would almost always tag along with his parents—unless he was absorbed in a good book. In that case, even as a teenager, he'd stay home and keep the fire stoked for his parents' return. Randy was the cliché boy under the covers with a flashlight. Many mornings, he'd wake with the house flashlight (batteries dead) in bed with him, having pushed it for one too many pages the night before. Even if television reception had been possible, the Morgenson household would have resisted. Radio, records, and newspapers were the main sources of news and entertainment, the *San Francisco Chronicle* and columnist Herb Caen being the family favorites.

In the winter, Randy would ice-skate on the pond at Curry Village and ski at Yosemite's ski area, Badger Pass, where his brother was an instructor and resident hot-dogger, who Randy looked up to and tried to keep up with.

During the mid-1950s, Larry was drafted to fight in the Korean War. While he was away, Randy and one of his best childhood friends, Bill Taylor, outgrew the "tame terrain" of what they coined "Badger Piss." The wooded glades and steeper slopes of the backcountry became their new playground.

NEPOTISM REPORTEDLY IS a major factor in securing choice positions in the national parks, and so it was in 1958, when 16-year-old Randy Morgenson applied for the coveted job of bicycle-stand attendant at Curry Village. The $1.35-an-hour job—his first—consisted of 28 hours per week renting and repairing bicycles, giving directions, and answering questions about the park. He was rehired the following summer for the same job at the same pay rate and, by age 17, he had saved enough money to buy his first car, a 1932 Ford Model B five-window coupe, for $200. By the end of that summer, he'd torn out the stock Ford engine and replaced it with a Cadillac engine. According to his friend Randy Rust, it "purred." For a time, it was his passion. "If he wasn't out in the woods somewhere," says Rust, "his head was either in a book or under the hood in their driveway."

In June of 1960, at 18, Randy took a job at the park's only gas station, where he serviced cars, repaired flats, sold batteries, and acted as a guide to the steady stream of visitors who were relentless in their barrage of questions regarding Yosemite. Randy had absorbed enough trivia from his father—the park's resident walking, talking Yosemite guidebook—to answer most queries with a flare, unexpected from a youth with a greasy rag hanging from his back pocket.

If someone asked directions to Snow Creek Falls in early June, he'd rattle off road directions, then probably suggest the Mirror Lake Trail, recommending, "Keep your eyes open for a heart-shaped leafy plant at ground level in the shady spots. Rub one of the leaves between your fingers and smell it for a surprise." The "surprise" was wild ginger.

As Randy began his senior year of high school, the Golden Age of big wall climbing in Yosemite was under way. Royal Robbins had pioneered a route up the 2,000-foot northwest face of Half Dome in 1957—the biggest wall that had ever been climbed at the time. Shortly thereafter, Warren Harding topped out on El Capitan's 3,000-foot Nose. Like many of the valley's youth, Randy occasionally loitered around Camp Four, which over time would become a mecca for climbers in search of what many call the vertical world of Yosemite. For Randy, it was simply another designated camping area in his backyard.

For Dana and Esther Morgenson, it was a source for concern. Randy had proven himself adept at all wilderness pursuits. He graduated from one to the next with the ease of a natural athlete. First it was ice skating at the ice rink, but then he decided that a remote lake was less crowded and afforded him more adventure and speed—the same way he'd progressed from controlled ski runs to the wild snow of the backcountry. Randy, like all the Yosemite kids, loved to climb around on the boulders and up the huge granite slabs along the edge of the valley floor. The Morgensons knew it was only a matter of time before they'd be holding their breath while watching him with binoculars—a fly on one of the very granite walls that they had long associated with the comforting safety of an unhurried life. Unlike the small-town kid who wants to go and discover the big city, Randy wanted to venture deeper and deeper into the wilderness. Day hikes became overnight adventures, and if he could find no one to go with him he had no qualms about going alone.

He became interested in stories of mountaineering and dreamed, not of the world-class rock climbing in his backyard, but of exploring deep into the mountains of exotic lands. Sometimes he would head out of the house with the goal of finding the most obscure spot to read, uninterrupted, a book or the newest *National Geographic* cover to cover. One of the more inspirational articles he read was about the famed Sherpa of Nepal, mountain people whose physiological makeup from living for centuries at the world's highest altitudes enabled them to travel effortlessly in the thinnest air on the planet and had made them favorite porters and guides for climbing expeditions. Randy marveled at the notion of traveling to the land of the Sherpa, but until then, the Sierra served as a training ground.

RANDY GRADUATED from Mariposa County High School, an hour's bus ride from his home, in June of 1961, ranked academically fifteenth out of forty-five graduates. He had excelled, especially in English and physical education, carrying straight A's through his high school career. Math, history, and science were B subjects. In sports he lettered in both football and basketball, but most telling was his elected position as senior class president, which friends attributed to his likable nature and way with words. Randy Rust remembers that he was a natural speaker and "comfortable talking about anything with anyone."

Much to the pleasure of his proud parents, Randy was accepted by Arizona State College in Flagstaff (renamed Northern Arizona University in 1966), where he declared his major as Recreation Land Management, a fairly new curriculum nationwide. But after a year and a half, the 21-year-old decided to take the spring semester off to work his first real job for the National Park Service, that of "ungraded laborer."

By all accounts, this was the first time Dana and Esther weren't pleased with Randy. Their elder son, Larry, had returned from the war and moved in with them, spending much of his time at local watering holes and often coming home drunk. Then, after vanishing for a few days, he appeared at the front door with a young woman he introduced as his wife. He'd met her at a bar, and after a brief love affair, they'd driven to Vegas and gotten married.

So, as one son was sinking into the depths of alcoholism, their intellectual son, for whom they had high hopes, was maintaining park trails with a pick and shovel.

It was to their great relief that Randy continued with school the following fall, explaining that he'd needed some time to clear his head. Truth be told, he didn't want to continue using his parents' money on an education that he wasn't excited about. His mind was in the mountains, and even then he knew that, unlike his father, working a job where he could "walk in the woods on the weekend" wouldn't be enough.

During the summer of 1963, Randy and his friend Bill Taylor headed south to Sequoia and Kings Canyon National Parks to hike a portion of the High Sierra Trail, which crossed the Sierra range east to west. On their last day in the backcountry, Bill reminded Randy that Randy was supposed to work at the theater in Yosemite for that evening's show. It was late morning and they were miles from their car; still, Randy paused at a particularly scenic overlook and marveled at the view—for nearly an hour. Randy continued to take his time, investigating trailside flower patches, pausing to photograph the crystalline stalactites dripping from sugar pine cones.

Randy's casual pace stressed Bill and he repeatedly reminded Randy about the time. Finally, Randy walked over to him and calmly said, "You're missing way too much by staring at that watch. Either throw it off this cliff or stop bothering me about being late."

"Randy did that to me a lot," says Taylor. "He reminded me to keep my priorities straight."

Indeed, Randy sauntered into the Yosemite theater just as the line of people were let in.

Bent on saving money, Randy took two more jobs from June to

September of 1963: he worked as a messenger delivering and collecting cash for the Curry Company, and he showed employee training films on the side, earning $300 a month.

In the fall of 1963, Randy returned to Arizona State College ready to hit the books. His first two years had been lackluster, with B's and C's the norm, and even a couple of D's.

Philosophy changed all that in his third year.

Introduction to Philosophy, American Philosophy, Critical Thinking, and Classic Piano all pocketed him A's. During this inspired time, he added books on Aristotle and Plato and other "great thinkers" to his bookshelf. But it was Confucius who probably best described the philosophical bent on wilderness that would last the rest of Randy's life: "Everything has its beauty but not everyone sees it."

Randy not only saw beauty in the smallest things, but also was captivated by their smallest details. He decided to spend the following summer in the high country. He wanted to put his life on his back, not unlike John Muir, and hike the crest of the Sierra without a schedule. Unhurried. Unhindered.

He informed his parents of his summer plans during Christmas break, which was an adventure in itself. Perhaps inspired by some great thinker, Randy attempted to return home to Yosemite by jumping a train. To test his mettle he left Flagstaff with no money. He didn't make it to the California state line. While his train was stopped at a rail yard, the cars were searched by a conductor, who discovered the unlikely hobo and kicked him off. He walked to the next town and called his parents, collect.

For the entire summer of 1964, beginning in Yosemite, Randy hiked the John Muir Trail south to Mount Whitney. Bill Taylor was one of the people he enlisted to hike in and resupply him with food caches along the way.

"Meeting the backcountry rangers on the trail," says Taylor, "made quite an impact on Randy. He did everything possible to stay in the mountains that summer. He didn't want to hike out if he could help it. Seeing the rangers along the trail, self-sufficient, with a cabin, was very romantic to Randy."

Whether it was the rangers, the high country, or just the best way to stay out there, Randy decided to apply for a backcountry ranger position for the following summer posthaste upon his return from the mountains, but not in Yosemite. The less crowded Sequoia and Kings Canyon to the south was the country he most enjoyed on his summer-long trek.

He couldn't have chosen a better time. The results of six back-country-use studies conducted over the previous twenty years had recently culminated in a landmark backcountry management plan for Sequoia and Kings Canyon. It was the early 1960s, and there was a new environmental movement that went beyond simply setting aside wilderness for future generations. These studies and others proved that wasn't enough. The land had to be looked after more closely than in the past. It had to be "managed," with a sensitivity for the wilds.

The management plan proposed an increase in backcountry rangers—which, since World War II, had numbered fewer than four rangers per summer season. Randy was on the cusp of a hiring movement that would triple the number of backcountry rangers in Sequoia and Kings Canyon.

DURING THE MONTHS after the summer of 1964, Randy decided college wasn't for him. He felt strongly that anything he was to learn on this planet would be taught to him by the mountains.

He told his friends that he'd learned more during those months in the high country than all his schooling up to that point, and he wanted to share what he'd seen and what he'd felt. But there was a dilemma. He couldn't talk openly about these aspirations with his parents because they were set in the belief that a college education was required to make a respectable living. They supported his love of wilderness wholeheartedly; his mother would say he "got that honest" from his father. If he wanted to make a life of the Park Service, wonderful. But the administrators—the superintendents, the chief rangers—had degrees.

Randy had a bit of the sixties in him and wanted what was then just beginning to be referred to as an "alternative" lifestyle. He wanted to create for himself a life where living came before a job. He didn't want to settle for one or two weeks of vacation a year. A desk job, whether in a suit and tie or a ranger uniform, was out of the question.

For advice he went to his family friend Ansel Adams, whom he'd assisted occasionally in his younger years. When Randy had first offered his services to the famous photographer, he was a young teen. He'd been worth his weight in gold during photography courses when he lugged Adams's heavy tripod and large- and medium-format cameras all over Yosemite.

As the years passed, Randy experimented with photography himself. While his father was bent on documenting the park's flowers and scenic vistas, Randy exhibited more artistic tendencies, which Adams observed as he reviewed his work.

Randy wrote Adams a letter in October 1964, explaining his predicament with school and career, and expressed his desire to use photography as a means to support himself while documenting his adventures in the High Sierra and eventually around the world.

"You make a very clear statement of your problems and I must tell you I have a very high opinion of your attitudes!" wrote Adams in response. "To see and to feel is supremely important, and to want to share your experiences is the hallmark of a truly civilized spirit!

"Photography, as a . . . profession, is a grim business—with terrific competition. Frankly, I advise most people to approach it as an avocation. Many of the greatest photographers were 'amateurs' in the sense that they did not make their living from the art. You have too fine a concept of the creative obligations to get yourself mixed up with the 'nuts and bolts' of the camera world."

Adams then invited Randy to come to his home in Carmel, California, to discuss the subject at length.

"I think I can help you much more in this way," Adams continued. "It is easy to write down ideas and suggestions, but you have a very definitive purpose in life (rare!) and I think I could help you most by just talking with you and exploring."

The dialogue between Randy and Adams from that meeting is unknown—but Randy did leave Carmel with a gift, one of Adams's classic wooden tripods and a 4-by-5 view camera. A few weeks later Randy dropped out of his fall semester courses, and applied for the job of seasonal backcountry park ranger on February 8, 1965.

On the application, under Special Qualifications, he wrote: "Entered public speaking contests in high school, and have been meeting the public and working with people all my life; have been backpacking through the Sierra covered by these 3 parks [Yosemite, Sequoia, Kings Canyon], plus some, for as long as I can remember."

Two months later, Randy was informed of an opening at Sequoia and Kings Canyon. He arrived at park headquarters at Ash Mountain on April 29. He was honored to serve in the Park Service and had proudly purchased the classic ensemble of olive green coat, gray shirt, dark green tie, and the traditional tan flat hat. The silver National Park Ranger badge represented something important. Just a few weeks earlier, he had been in Arizona, majoring in outdoor recreation. Now he was going to live it, to actually get paid to go camping in the mountains.

On May 1, Randy reported for duty at the parks' vehicle entrance kiosk, not far from Ash Mountain, where he would work for a few weeks before being dropped into the backcountry.

The entrance kiosk, or check-in station, was the hub of activity at this, the parks' southernmost entrance. With the passing of the Wilderness Act in 1964, the parks had experienced a slight increase in traffic. Still, check-in station duties had changed little since the early 1940s, when Gordon Wallace, a ranger in Sequoia from 1935 to 1947, worked inside the same rock-walled building. Wallace recounted his duties in his memoir, *My Ranger Years*:

> Not only must all traffic, local as well as tourist, stop here and make its business known, but the station also served as the clearing house for all the trivialities, petty bothers and errands, information of all kinds, and amenities of daily life. The park ranger on duty at the checking station was the pivot of this life. . . . Besides the locals, I have seen and talked to many others—people

who came from the forty-eight states as well as from other parts of the world.

From years working customer service jobs in Yosemite, Randy knew that a smile combined with enthusiastic local knowledge went a long way when dealing with the public. Randy's performance at the check-in station prompted accolades, as acknowledged by a letter written to the park's superintendent, John M. Davis, on November 5, 1965, in reference to a family's encounter with Randy on June 8, 1965:

Dear Sir,

On behalf of the attitudes promoted in Sequoia National Forest [sic], I must comment that it's wonderful to know that for those traveling throughout our great country, there are individuals and systems set up to further interests and establish atmospheres of enjoyment for all who wish to grasp the beauties of America.

In particular, I refer you to Mr. Randy Morgenson, a ranger who attended the check-in station. . . . His brilliant character, sparkling personality and cheerful smile both entering and leaving Sequoia left an impression my family and myself will never forget and I'm sure made the long trips for the many who passed through Sequoia that day bearable ones. . . . We appreciate it very much. Thank you.

Please give Randy our sincerest regards and the enclosed picture we took of him on our way through.

Gratefully yours,
Mrs. A. Wayne Ingard
Moscow, Idaho

Superintendent Davis forwarded the letter and photograph to Randy, and responded to Mrs. Ingard with his own letter, which stated, in part, "Service to Park visitors is one of our primary functions, and we are always happy to hear that this important service is being carried out cheerfully and courteously."

Mrs. Ingard's glowing letter was the first of dozens that would eventually be filed in Randy's meticulous archives in the attic of his home. None of these letters would be included in his government employee personnel file.

Even though Randy performed his duties admirably at the entrance station, it was not the reason he had joined the Park Service. Like his predecessor Gordon Wallace, Randy longed for the backcountry. It was a calling that had been eating at him nonstop since he'd hiked the John Muir Trail the summer before. He wanted to get far and away from the cars and blacktop of the parks' most traveled routes and sites: the General Sherman Tree—with its 36.5-foot-diameter base, the biggest tree in the world—and the myriad quick roadside hikes that could be enjoyed by anybody with a few hours to spare while passing through.

After six weeks of inhaling exhaust fumes at the parks' entrance, Randy helped load what appeared to be supplies for a small expedition into the belly of the parks' helicopter. His first season as a backcountry ranger was about to begin.

As the pilot gained altitude, the cabins and roads at Ash Mountain became a distant memory. On a northeast flight path, the helicopter skimmed the granite walls of Moro Rock, taking a wide arc around Giant Forest, where groves of the world's largest trees seemed toylike in comparison with the serrated teeth of the snow-clad Sierra Crest that filled the horizon to the east. Following the routes of rivers and streams, the pilot weaved into the high country as his wide-eyed passenger spun around to absorb every geographic feature. It was Randy's first bird's-eye view of a land he would describe to his mother as Eden.

Gordon Wallace had taken a similar eastern route into the Sierra wilderness some thirty years earlier, though his summer ranger supplies had been transported by a string of mules. Wallace's stock had grazed freely in any and all meadows; Randy's rotor-powered steed was deemed less invasive to such meadows, despite the noise.

"Do not come and roam here unless you are willing to be enslaved by its charms," warned Wallace in his memoir of his ranger years. "Its beauty and peace and harmony will entrance you. Once it has you in its power, it will never release you the rest of your days."

By the time Randy jumped out of the helicopter onto the gravel shore of Middle Rae Lake, it was already too late. The spell had been cast.

CHAPTER THREE

INTO THE HIGH COUNTRY

I went to the woods because I wished to live
deliberately, to front only the essential facts of life, and
see if I could not learn what it had to teach, and not,
when I came to die, discover that I had not lived.

—*Henry David Thoreau,* Walden

Only this simple everyday living and wilderness
wandering seems natural and real, the other world,
more like something read, not at all related to reality as
I know it.

—*Randy Morgenson, Charlotte Lake, 1966*

WHEN RANDY MORGENSON hopped off that helicopter near the shores of the Rae Lakes on July 12, 1965, he landed in a new era of wilderness. The early environmental movement had long fought for the idea of protecting the wilds, not exploiting them. Now, with the passing of the Wilderness Act the year before, the National Park Service was struggling to balance two wholly conflicting philosophies mandated by the new law—preservation and use. Even pre–Wilderness Act, Sequoia and Kings Canyon had implemented grazing restrictions in certain areas where heavy use, if continued, would have turned mountain meadows into dirt fields. Other areas, the Rae Lakes in particular, had been so heavily used by campers that dead and down wood that could be burned as firewood was almost depleted. In cases such as this, camping and grazing of stock was limited to one night, and a recent invention for cooking—the backpacker stove—was encouraged. Without these and other controls, it was predicted, the High Sierra wilderness would never recover.

Referencing a range of ecological studies, the parks' scientists compiled a backcountry management plan in 1960 that outlined ever-increasing populations. They proposed a set of experimental rules and regulations that, if adhered to, would theoretically save the backcountry from becoming another frontcountry.

Randy represented a new generation of clean-shaven and uniformed rangers with military-cropped haircuts who, like military grunts, were stationed on the front lines but, as seasonal employees, held the lowest rank. Their challenge was to introduce this way of thinking to a cast of backpackers, fishermen, horsemen, and climbers, who weren't always receptive to new ideas.

In young Ranger Randy, the Park Service had been delivered the perfect foot soldier, though his gentle nature made him more of an archangel crusading in a green uniform. He already considered the High Sierra his church; the backcountry management plan became his bible. The report read like scripture to Randy, warning of an impending doomsday and often citing his childhood home, Yosemite, as an example of what could occur. Here, in the backcountry of Sequoia and Kings Canyon, man's presence had not yet dismembered Mother Wilderness—but she was barely holding on.

Armageddon was upon them.

Besides his academic knowledge of that report and a genuine desire to protect his beloved wilderness from the proverbial fires of hell, Randy had brought with him an innate love and enthusiasm for the Sierra as well as the ability to survive in its wilds. The plants and the animals were his kindred spirits; the geology and waterways were his temples. But he didn't know first aid or CPR. He wore no sidearm, carried no handcuffs. Disarming, much less defending himself from, an armed suspect was the stuff of movies, not his reality. The skills required to lower an injured climber off a precarious cliff or rescue a drowning hiker from swift whitewater had not been taught in ranger training because there was no formal training for seasonal backcountry rangers at Sequoia and Kings Canyon.

His job was to hike the trails and "spread the gospel" to as many visitors as possible. He issued fire permits, picked up trash, hung Mountain Manners signs, naturalized campsites, and if there was an emergency—medical, forest fire, whatever—he was to tend to the situation as best he could and radio for assistance. In 1965, he knew none of the skills that would become second nature as he traveled down the high and lonely path of these parks' most trained—some would call them elite—backcountry rangers.

Despite his fit, but certainly not commanding, 5-foot-8-inch, 140-pound frame, Randy was, in these parts, the law of the land. Add youth to his stature, however, and the National Park Service patch on his shoulder and silver badge on his chest gave him little more law enforcement presence than an Eagle Scout at a bank robbery. But that didn't mean he didn't take the rangers' motto seriously. Each morning he pinned the National Park Ranger badge above his left chest pocket; he was prepared to "protect the people from the park, and the park from the people." It was his mantra.

He soon learned, however, that his main duty that summer was to collect garbage—gunnysacks full of it. The second-most-prevalent chore was cleaning up "improved campsites," which meant tearing down the log-and-granite dining tables and kitchen areas sheltered behind rock-wall windbreaks. Fire pits were his nemesis, engineering feats he deemed "fire castles" for their sheer immensity. They often came complete with iron grates that campers hid in nearby hollow logs or hung from trees when they left the mountains. Generations of families had been coming to these spots for years—sometimes kicking out other campers who were squatting on *their* campsite. Imagine their surprise when they couldn't find "their" campsite and a young, mus-

tached Ranger Randy materialized out of the woods to explain that the
area was being "naturalized."

"Natural-what? I just want to know where my fireplace is!"

It was predictable. The parks' management plan had a section entitled "Wilderness Protection vs. Personal Freedom," in which was written, "Oldtime use of wilderness was completely free of restrictions.
Wilderness explorers could hunt and fish without limit, cut down trees
at will, camp, make fires and graze their stock anywhere. The tradition
of personal freedom in wilderness dies hard. . . . But when human populations expand they become subject to the biological limitations that
govern other dense populations: the greater the number of individuals
the greater the loss of individual freedom."

Translated: "Sorry, sir, the fireplace your grandfather built with
your father has been obliterated, but I replaced it with this highly func-
tional, less obtrusive fire ring that's—yes, sir, I realize it's quite small,
but it will still provide plenty of warmth and cooking surface, not to
mention you won't have to burn an entire tree each time you light it.
By the way, you won't be needing that ax. The new regulations allow
only foraging for deadfall on the ground. Oh, and please don't cut pine
boughs for your bed—that's illegal now as well. Have a nice day."

Randy, who was neither so blunt nor so stiff, strove to respect past
freedoms, introducing the new rules and regulations to more than
1,200 park visitors in his patrol area that season without hearing a
complaint. The only citation he issued was to a backpacker who had
brought his dog with him, which led to a discussion about the difference between national parks and the national forests bordering the
parks, which are managed by a much looser set of use regulations. That
first season was devoid of any major emergencies: Randy treated one
person for blisters, and a dehydrated girl who felt sick merely needed
to force down water. He destroyed seventy-five oversized fire pits and
collected thirteen gunnysacks of garbage that were hauled out of the
mountains by mules. As the summer progressed, he earned his reputation as a devoted and diplomatic workhorse who once hiked 16 round-
trip miles to tear down a haphazard community of campsites that he'd

heard was destroying the serenity of a remote lake. Exhausted after hours of moving rock and logs, he embarked on the 8-mile return to his station and discovered en route one of the Sierra's legendary can dumps—a rusting midden that couldn't be passed by. After loading his pack with 50 pounds or more of glass and cans, he returned home well after dark to collapse in his sleeping bag.

He lived in the spartan accommodations of a tent on the shore of Middle Rae Lake and recorded his simplified life with the romanticized pen one might expect from an inspired 23-year-old truant from society who had been raised on a diet of nature writers such as Ralph Waldo Emerson, Aldo Leopold, and Henry David Thoreau. "There is a low plant that grows profusely everywhere, composed of a 'cup' of several leaves pointing nearly straight up," Randy wrote after some afternoon rain showers. "The whole is maybe half an inch high, and they form carpets that one could take for a meadow. Whenever it rains a large drop of water collects in the bottom of this cup and glimmers like the brightest diamond in a green rosette when the sun comes out . . . the most brilliant diamond one could imagine."

After one patrol, he returned to his station, where "the evening alpenglow on the peaks filled me with a feeling of bigness inside," he wrote. "As I rounded the final edge of lower Rae the jumping fish dotted the lake with their rings. So still was the air I could hear the splashing . . . and as they jumped clear of the water I could see momentary flashes of silver bodies. Descending the final slope to my cabin, looking out over Arrowhead Lake and toward Pinchot Pass in the diffused pinkish light, I felt positively exhilarated. I know exactly how Henry Thoreau felt when running home after the rain. 'Grow wild according to your nature, like these brakes and sedges which will never become English hay, let the thunder rumble.'"

Indeed, Middle Rae Lake was his Walden Pond; the surrounding peaks, basins, and meadows were his Sand County.

There was something different about staying in one general area that Randy's previous summer's hike on the John Muir Trail had not revealed to him, a satisfying sense of ownership that came with the job: not a selfish, territorial bent, but more the pride a homeowner feels for his property. With that sentiment came a respect for his endearing neighbors—the marmot family that had taken up residence in a burrow near his cabin's doorstep and the rosy finches and Clark's nutcrackers that vied for his attention as he strolled down the trails.

The fragile mountain meadows became Randy's personal cause, no doubt impassioned by childhood walks with his father and brother. If a packer grazed his mules in a closed meadow or a poorly informed backpacker pitched his tent on anything green instead of on gravel, it was as if they had desecrated Randy's yard—the church's gardens.

His supervisor, a well-liked ranger by the name of Dick McLaren, gave Randy a line of advice to which he would adhere for the rest of his career: "The best way to teach the public isn't with a citation, it's with communication." And so Randy would offer to help move an ill-placed camp or catch an uncooperative mule in a wet meadow, and gently explain the reasons behind the rules—sometimes to the packer, sometimes to the mule, to the amusement of the packer. A story that would become legendary in Sequoia and Kings Canyon was about the backpacker who asked Randy the name of a tiny flower he had pitched his tent upon. Randy apologized and told the backpacker that he knew only the flower's "book name." He explained that he hadn't figured out how to ask the flower its real name, but thanked the backpacker for his interest. The hiker likely never pitched his tent again without carefully checking what was underneath it.

But the living flowers, grasses, and animals weren't all that tugged at Randy's heartstrings. Even the granite peaks—cast in a surreal glow each morning and night—hypnotized him with their sublime, quiet beauty and mystery. Among these high crags were secret passageways, long forgotten or never explored, that called out to him. After staring at one such cleft for more than two months, he devoted one of his days off to satisfying his curiosity. With some difficult scrambling and climbing, he reached the crux, which was a doorway into a hidden basin enclosed by an amphitheater of stone, where water flowed literally from solid rock.

As he crossed the threshold of the notch, it was as if the mountains were sharing a verdant secret with him. He described it as "some of the most beautiful country in this area, perhaps because it is pure—untouched, untrammeled, and unlittered." He explored the shores of silent glacial tarns, finding no other footprints. The flowers grew as they should at these heights, where the soil's nutrients had gathered, sustaining them in "small patches or tufts between the boulders," without fear of being plucked or smashed by a hiker or eaten by a mule. There were no blackened fire pits or piles of rusting cans, though there was a flat spot above the meadow that had been someone's barely perceptible sleeping spot. The haven he'd been drawn to was, according to Randy, "rich country," symbolizing not only the past but also what he hoped would be the future for these mountains.

AS RANDY RELAXED into the daily regimen of life as a backcountry ranger, Dana and Esther Morgenson were increasingly anxious back in Yosemite. They weren't concerned for his safety in the mountains— they were confident he could handle anything the Sierra might throw at him.

They were, however, worried about the rumblings of a draft. On July 9, 1965, just three days before Randy was airlifted into the backcountry, President Johnson acknowledged in a news conference that his administration was considering a call-up of reservists and expanding current military draft quotas. Randy was of age, and Dana and Esther knew that no amount of wilderness could shield him from the Selective Service and Vietnam. The Morgenson family had seen what war could do to a person's spirit. Larry, whom Randy had once looked up to as an artistic and talented storyteller, a tireless skier, an older brother with worldly aspirations, had atrophied after the Korean War into living his life within the constraints of a bottle. The drink helped curb what would later be called post-traumatic stress. Regardless of the reasons behind Larry's uninspired life, family and friends marked the beginning of the decline with his military service. Even knowing this, Randy had told his parents that he would serve his country if he was drafted. He "wouldn't like it," he said, but if he was called, he would go.

Toward the end of the season, Dana and Bill Taylor—Randy's childhood friend—hiked into the backcountry for a visit and were surprised to see how much weight he had lost. It was impossible for a foot ranger not to lose weight; he simply could not consume enough calories at altitude, especially with a canned-food diet. They brought with them homemade cookies from Esther, which Randy rationed sparingly after meals.

Seeking his father's expertise, Randy told him of the flowers that had appeared like diamonds after the rain. Dana instantly recognized the description as bilberry, but he and Randy hiked to the spot to confirm. The conversation, as it often did, segued into school.

Bill Taylor thought that with the threat of a draft, Randy was crazy even to consider not going back to school. Full-time students were eligible for deferment, a no-brainer to Bill. Dana expressed his concerns as well.

If Randy went back to school in the fall and spring, he could come back to the high country the following summer. That wouldn't be an option if he were to be drafted. He agreed to think about it.

Back in Yosemite, Dana confided his concerns to Randy's friend Nancy Williams, a young woman who worked with Dana in the Curry Company's accounting department. Dana expressed to her his disappointment in Randy for not continuing his education and his worry that he was exposing himself to the draft. But Nancy understood that "Randy was answering a higher calling." She describes it as an irresistible pull, like Jack London's "call of the wild." "I think Randy had a distinct purpose in life," Williams says, "and back then, he wasn't exactly sure what that purpose was. He just followed his heart, which wasn't in the classroom. The *mountains* were his classroom."

Such idealistic reasoning provided Dana and Esther little respite from their worries. War, they knew, was not their son's calling—he wasn't programmed for it. Before he'd left for the mountains, they had urged him to continue with his education. In the mountains, Randy reasoned, that was exactly what he was doing. When he wasn't writing in his logbook or practicing with the camera Ansel Adams had given him, he was memorizing the backcountry management plan.

As summer edged toward fall, Randy made it a point to speak with everybody he encountered. His knowledge and charm led people to invite him for dinner at their camps, and he reciprocated by inviting backpackers into his tiny yet cozy cabin for tea when a rainstorm passed overhead. Despite his comfort with solitude, he was extremely social and could dive into conversation and not come up for air for hours. On one long patrol to Upper Basin, he met a man and his daughter atop Pinchot Pass and spent some time with them, chatting about the Sierra. Afterward, Randy turned to hike down into Marjorie Lake Basin, toward Bench Lake. The father, obviously impressed by the young ranger, told him, "I hope this is your career—we need you."

"I was pleased," wrote Randy in his logbook, "that he felt I was a credit to the service."

FOR WHATEVER REASON—to avoid the war or to please his parents— Randy was back at Arizona State College in Flagstaff the following fall. He carried with him the memories of an enchanted summer, and a folder that bore the quote:

Wilderness

An area where the earth and its community of life are untrammeled by man, where man himself is a visitor who does not remain.

—Howard Zahniser

Perhaps it was a tribute to Zahniser, a former executive director of The Wilderness Society, as well as the author of the Wilderness Act. He had died four months before President Johnson signed the act into law on September 3, 1964. Or maybe it just kept Randy's mind in the right place as he studied cultural linguistics, religious philosophies, and Asian cultures and philosophy, all classes that played into his longstanding dream of visiting the tallest mountains in the world: the Himalayas.

Not unlike the military, the Peace Corps used romantic photographs of exotic locales to entice potential volunteers and recruits. Such photos of Asia, and in particular the Himalayas, struck a chord when Randy happened by a Peace Corps booth at his school, where, as in many other college towns and campuses, Peace Corps recruiters shared the sidewalks with Army, Navy, Air Force, and Marine recruiters. In 1966, the Peace Corps was considered either an honorable exemption from the draft or, as Richard Nixon put it, a "haven for draft dodgers." The crowds gathered around the Peace Corps recruiters told the story of that era.

As with many things in his life, Randy didn't consider being drawn to the Peace Corps to be merely a chance encounter. He filled out an application, requesting Asia as his top choice. There were no guarantees, but he made it clear that the Himalaya region was his dream assignment—another beckoning doorway that seemed to be leading him down a specific path in life.

"The master intended for me a life in the wilderness, a life of awareness and discovery of the forces of nature and humanity. A life . . . that carries me toward more entire manhood, and perhaps one that brings some of this into the rest of the world, counterbalancing some of the forces that presently carry us along."

Randy wrote this in a letter to his parents while looking out from the terrace of the mud house he had been assigned in the village of Golapangri, in the Maharashtra region of India. It had been more than a year since the Peace Corps acceptance letter was delivered to him by mule at the Charlotte Lake station during his second year as a backcountry ranger. His journey from that point forward continued like a fable.

RANDY HAD BARELY STRAYED from Yosemite's granite womb when he boarded a jet and set out on a pilgrimage to the abode of the gods, the land of the Sherpa who climbed the mountains of the Himalayas with reverence and awe. To gaze upon those majestic peaks and walk even in their shadows was the ultimate treasure he sought when he joined the Peace Corps. Any assignment, he felt, would be worth the reward.

For two years he lived in a small village nearly 2,000 miles from the mountains of his dreams. His vistas were dry and dusty farmland, void of anything green, and flat for as far as the eye could see. The Himalayas were "over there" somewhere beyond the horizon, where the sweltering 115-degree heat distorted the view. He would awaken each morning and watch the village come to life: women bringing the day's water home from a central well with jugs balanced on their heads, bullock carts bouncing off toward the fields, smoke from cook fires, and neighbors chatting over mud walls. "It seemed a thousand or two thousand years ago," he wrote. Everything was exotic, from the colorful open-air markets, always with Indian music blaring "to the point of distortion" from unseen speakers, to the slow, rural pace.

While Randy taught the farmers "Western" agriculture techniques, the Indians taught him their religion. He came to understand the prayer rituals at the village temple, the daily offerings at family shrines, the deities—more so, he thought, than he might have learned had he stayed at the university his parents wished him to attend.

One day, Randy's Indian friend Limbaji explained how everybody in the village thought he was a Christian. Randy, who had erected a Christmas tree that December in his mud home, replied that he did not consider himself a Christian.

However: "Your people are Hindu, my people are Christian; you are an Indian, I am an American; your skin is dark, my skin is white," he said. He held his pale arm against Limbaji's dark skin. "What is the difference?" asked Randy. "There is no difference—we are the same."

At this, Limbaji grinned widely and reached for a stone. "But this is what different religions mean," he said, placing the stone on the ground. "God is for all men, he is always the same. There is only one. And all men finally go to the same God." He drew lines toward the stone in the dust. "But there are different roads."

From the dry seasons to the monsoons, Randy put his 720 hours of training to the test as one of fifty individuals in the Peace Corps India Food Production Project. By the end of two years, Randy and the other volunteers had shown the Indian farmers how to double, sometimes triple, the yields of their crops. It seemed they had, after many road-blocks, succeeded in their quest. Not long before Randy left, he asked one of the farmers whom he'd worked especially closely with if he intended to continue farming the land as he had been taught.

The farmer, with a cheery disposition, said no, they wouldn't. Once the volunteers left, he explained, most of the farmers would go back to their old ways.

Randy couldn't believe his ears. "Why?" he asked.

"Because," said the farmer, "that is how you farm in America. This is how we farm in India."

Dumbfounded, Randy packed his bags and traveled down the roads of Eastern religions in Nepal, Thailand, Cambodia, China, and Japan. In the religious melting pot of Kathmandu, he began to favor aspects of both Hinduism and Buddhism. In Bangkok, he explored hedonism; it had been a long time since he'd been in the company of a woman. In Japan he was drawn to the meditative contemplations of Zen. But it was in the Himalayas that he experienced his greatest pleasures and felt most at home.

Randy enrolled in a monthlong guide school taught by Sherpas to learn technical mountaineering skills and expedition planning. When he completed the course, the school's head instructor, Wangdhi Sherpa, wrote a letter of recommendation in broken English, stating that Randy Morgenson "is keenly interested in the mountaineering and he has proved that his climbing tactics in rock and high altitude during the course. He is very cheerful all the time and good discipline among the peoples. We have no doubt he is a good mountaineering in the future."

Within weeks of finishing the course, Randy organized his own expedition to climb Hanuman Tibba, a 19,450-foot peak named after a Hindu god. It was reportedly only the third or fourth ascent (there was some dispute about one of the claims), which was an attraction, but first and foremost, the mountain was beautiful. He hired a high-altitude porter and two Sherpa guides, and climbed successively for eight days to establish a high camp. He experienced near-vertical slopes where the front-points of his crampons were all that were in contact with the mountain and there was nothing but air beneath his heels; he felt the sickening sensation of dropping suddenly as a snow bridge settled while he crossed a crevasse; and he understood the satisfying "thunk" an ice ax makes when it is placed solidly in "good snow." He learned to breathe and walk at altitude, to build anchors in the snow and rocks, and to work as a rope team in dangerous terrain and playfully torment his rope mates in safe terrain. He came to appreciate "bed tea" served by the Sherpas, and made it a point to awaken early one morning in order to return the favor to his bewildered crew, who had never been served tea by a Westerner, especially a Westerner paying for their services.

Randy summited the peak at 9:15 A.M. in June 1969 after a 3:30 A.M. start. The view was completely obscured by clouds.

Since he had first read about the Himalayas, "I've wanted to enter this world," Randy wrote his parents, "to live some time among the higher peaks, surrounded by ice and snow and deep blue sky only . . . a silent world. An intense world." He continued to trek through the Himalayas, visiting Everest's base camp and climbing a handful of peaks that approached 20,000 feet, always with just a few porters and Sherpas, for all of whom Randy prepared and served tea as a sign of respect.

"Now I've really become expedition minded," he wrote his parents. "I have thoughts about doing this sort of mountaineering again, beginning with winter mountaineering in the Sierra, and including vague intentions of returning to the Himalaya. Oh there are so many mountains: Alaska, the Andes, the Rockies, and Cascades, and yes, even the Alps. And such a late start I am getting.

"How wonderful to wander among virgin hills! I suppose whiteness is a symbol of purity (skin color being an exception) and how pure I found that world. As you've heard me say many times, the mountains are my life. Without them I am nothing. They are perhaps the only reality I know. They are my guru. If I am to learn anything in life, I will learn it there."

After three and a half years, Randy flew home. During the drive from the San Francisco airport, he was blessed with a clear day, able to look east toward the white-tipped spires of "his" mountains—the same "snowy saw-teeth" of the Sierra Nevada that captivated Spanish explorers when they sailed into San Francisco Bay in the mid-1500s. As he drew closer, he felt in his heart the pull of the high country. And so the moral of his fabled travels read like Santiago's, the boy in Paulo Coelho's *The Alchemist*: Randy had traveled around the world in search of treasure and came home to find it in his own backyard.

He placed the cherished letter from Wangdhi Sherpa into a box and filed his memories neatly, as one does with memories from great journeys. And then he made the telephone call to his old boss at Sequoia and Kings Canyon and held his breath after inquiring whether there was still room on his backcountry ranger staff.

"When can you start?" asked the district ranger.

AFTER HIS SABBATICAL overseas, Randy, at age 28, was assigned to deep, dark LeConte Canyon, where every day he awoke and looked to the sky, usually from his sleeping bag, beyond the tops of the lodgepole and white-bark pine to the granite spire of Mount Langile. As guardian of LeConte Canyon, Langile was first to feel the warmth of the sun each morning and last to bathe in its glory from the west come sunset. Randy tuned in to these and other cycles, noting that after the third week in July, the hermit thrush often stopped singing; that here in the canyon's bottom the robin's song was heard, but above 10,000 feet it scarcely, if ever, used its voice.

His job hadn't changed since 1966; neither had the physical attributes of the high country. But the spirituality of the place had shifted noticeably since his travels in the East. LeConte Canyon was no longer just a wooded canyon with sheer walls and a melodic rushing river. It was a massive meditation garden, the antithesis of the "superorderly" domesticated Japanese gardens that were "clipped, trimmed, and cleaned" to the point of sterility. In Kyoto, Randy had watched temple gardeners sweeping the dirt beneath trees. "No sooner does a leaf fall," wrote Randy in his diary, "than it is swept away and burned."

Randy preferred what he described as the "shaggy wildness of nature untended." He was "nourished" by the chaotic glacial rubble of the Sierra, where rotting tree bark and fallen pine needles obscured the lesser-traveled footpaths. There were no bonsai gardeners sculpting their human vision. Instead, there was the unpredictable, gusting winds that pummeled altitude-stunted pines to the finest artistic expression.

The influences of the religious roads his Indian friend Limbaji had spoken of—Hinduism and Zen Buddhism in particular—were synthesized here on his own chosen path, where his bible was still the Sequoia and Kings Canyon Backcountry Management Plan. When he tore down fire pits, he referenced in his logbooks for years to come that he was doing the work of Shiva, the Hindu "destroyer" deity who, appropriately, possessed contradicting powers—the ability both to destroy and to restore. He carefully placed the rocks in rivers to wash away burn scars, or buried them halfway in the duff of the forests, or carried them great distances from popular campsites to discourage campers from rebuilding the fire pits.

On September 8, 1971, Randy scrambled to the difficult summit of 13,034-foot Mount Solomon and wrote in the peak register: "We are the greatest bulldozers to walk erect. Will we ever permit, in a small place as here, Mother Nature—truly our Mother—to do her thing, undisturbed and unmarred? Will we ever be content to play a passively observant role in the universe, and leave off this unceasing activity? I don't wish man in control of the universe. I wish nature in control, and man playing only his just role as one of its inhabitants. I want every blade of grass standing naturally, as it was when pushed through the soil with Spring vigor. I want the stones and gravel left in the Autumn as Spring meltwater left them. Only these natural places, apart from my tracks, give me joy, exhilaration, understanding. What humanity I have has come from my relations with these mountains."

Such was the naturalist bent Randy conveyed to the public as he patrolled these remote mountains. But the outdoor recreation boom of the 1970s was upon these mountains, and with it came a crop of nontraditional visitors to the national parks. Even though miles of rugged wilderness separated his mountain paradise from civilization, the role of Randy—and all rangers—was on the verge of a drastic change of focus from gentle, approachable naturalist to law enforcer.

The catalyst in this movement probably began with what has come to be known as the Yosemite Riots. It was the Fourth of July 1970, and between 500 and 700 youths gathered to whoop it up at Stoneman Meadow, not far from Randy's childhood home. In contrast, that same day Randy patrolled 11 miles from LeConte Canyon to Dusy Basin and back, during which he saw only a dozen camps and thirty-four backpackers. "Quiet for the 4th," he wrote in his logbook. About the time he settled down for a simple dinner, a few rangers went into Stoneman Meadow's crowd of so-called hippies and tried to get them to disperse, explaining how they were damaging the meadow. When nobody budged, some mounted rangers announced a mandatory curfew. Details are sketchy from this point forward, but the youths—now allegedly a stoned and drunk mob—considered the curfew a challenge and stood their ground.

Then, "rangers, fireguards, and anyone else who could reasonably be put on a horse or asked to walk into the crowd, did," says one ranger. They had "no riot training. No nothing. They got thrown out of the meadow immediately after a brief skirmish where rocks and bottles were thrown."

The rangers, up against their first major civil disobedience encounter, regrouped and were joined by local sheriff's deputies for "special emergency assistance." By the time the night was over, nearly 200 youths had been taken into custody. The national media had a field day, and a recurring theme publicized the need for rangers to be better trained in law enforcement tactics and crowd control. The national parks had lost forever their identity as wilderness sanctuaries. The Granite Womb, it appeared, was not immune to urban crowds and violence.

As Randy continued his relatively quiet, oftentimes meditative, existence deep in the backcountry of Sequoia and Kings Canyon, the National Park Service geared up to avoid such a fiasco in the future.

In the wake of the riots, the Department of the Interior allocated to Yosemite a substantial budget to handpick and/or recruit a cadre of about fifteen rangers, many of whom had law enforcement backgrounds or special skills that might prove helpful in dealing with the youthful element frequenting the parks in the 1970s. Once at Yosemite, this group received special training in everything a modern-day ranger might require, from search-and-rescue tactics in the backcountry to emergency medical training to law enforcement tactics—physical, verbal, and psychological.

One objective was to create a nucleus of the best rangers to ever wear an NPS badge. This almost exclusively male group strived to climb better, ski better, provide the best visitor services, be the best emergency medical technicians, the best pistol shots—you name it, their goal was nothing less than excellence. "We wanted to deal humanely with these nontraditional visitors who frequented Yosemite in the early 1970s," says Rick Smith, a seasonal ranger who was recruited from within Yosemite's staff. "We were as good with a 5-year-old on his or her first visit to Yosemite as we were with a young person who came to the park to smoke his or her first joint."

In time this group came to be known as the Yosemite Mafia, and its influence resonated throughout the agency. "They were an incredibly talented group of people who, by force of personality and example, raised the bar on professionalization in every aspect of ranger work," says one veteran ranger who worked with many of the original recruits. But some rangers weren't excited about this new über-ranger mentality. The old guard continued to cling to the image of the friendly jack-of-all-trades ranger whose skill came from some mystical osmosis with the wilderness. The Randy Morgensons of the NPS had little interest in law enforcement.

Most of the Yosemite recruits became dedicated lifers—company men, so to speak—who worked their way up the ranks to hold positions at the highest levels of the National Park Service, from subdistrict rangers to district rangers to chief rangers to park superintendents— all the way to Washington, D.C., and the Department of the Interior.

Back in the early 1970s, the Yosemite Mafia tended to expand its talents by recruiting rangers with promise who would then hire and train other rangers who would then transfer to different parks. There was no grand design or mission, but as a result of these recruiting practices, the nation became permeated with a staff of top-notch rangers capable of handling whatever the public and the parks could dish out.

It wouldn't be long before Randy Morgenson's name came up for consideration.

CHAPTER FOUR

THE SEARCH

The map is not the territory.
                    —*Alfred Korzybski, 1931*

Lake Basin . . . I feel I could spend my life here.
                    —*Randy Morgenson, 1995*

IT WAS IRONIC but not unusual that some of the backcountry rangers gathered at the Bench Lake station on July 24, 1996, had said goodbye to each other a few weeks earlier with the casual parting statement, "See you at the SAR."

Others parted with "See you at the big one." All the rangers knew, even before they were flown into their duty stations, that search-and-rescue operations were inevitable. Despite potentially tragic outcomes, a search-and-rescue operation was still a ranger reunion—a sort of morbid social gathering where they steeled themselves against emotional ties with their fellow humans, usually park visitors who were missing, injured, or in peril or had already met their end. In that case, the word "rescue" becomes "recovery," synonymous with "body." Those in the business of search and rescue say there's only one thing that compares with the emotional strain of searching for a child, and that's searching for someone you know and care about. A recovery operation for either is without argument the most dreaded aspect of a ranger's job.

Both Randy Coffman and Sandy Graban had been summoned to such a tragedy not far from the Bench Lake ranger station in the summer of 1991. A 17-year-old girl had succumbed to probable high-altitude pulmonary edema on the last day of a backpacking trip with her family. The heartbreaking story, recounted in a case incident report, told of the girl's demise—her labored breathing, unsuccessful attempts to verbalize for help, nearly an hour of CPR—and the anguish of her parents. The deceased girl's mother stayed with her while her father and sister hiked out of the mountains over Taboose Pass. Thirteen hours later they reached the sheriff in Independence, who contacted the park's dispatcher, who notified District Ranger Coffman at home. At 9:30 P.M. it was Coffman's unpleasant duty to call the deceased girl's father at a hotel, both to inform him of the recovery plan and to lend a sympathetic ear.

At first light the following morning, Coffman was flown into the backcountry and met Graban, who at the time was the Bench Lake ranger. Together, they rendezvoused with then–Subdistrict Ranger Alden Nash at the family's campsite near the shores of Bench Lake. The rangers provided comfort to the grief-stricken mother while respectfully investigating the scene. Nash, Coffman, and the helicopter crew then carried the girl nearly a mile to a suitable landing zone, where the parks' helicopter transported her body out of the mountains.

Randy Morgenson himself had responded to equally tragic calls for climbers who had fallen, in some cases hundreds of feet. These deaths were precipitated by loose rocks, a patch of ice, or a momentary lapse of attention. So violent were some of these incidents that clothes and even shoes were ripped off.

The parks' rangers knew well what granite can do to a human body, and the merciful, albeit macabre, reality for rescuers was often the unrecognizable state of the victim, who sometimes appeared more like a mutilated deer hit by a truck than a human being. That's how they dealt with it. Mechanically. Impervious to the blood and thankful when there was no face to attach to the memory.

Coffman, Graban, George Durkee, Lo Lyness, and Rick Sanger had all witnessed death at some point in their careers. They knew what could happen in these mountains.

It was this unknown that was most troubling during the Morgenson SAR, an ambiguous voice that whispered into the ears of these rangers an incessant list of worst-case scenarios. A loose rock had pinned Randy; a rock slide had buried him; an icy log had caused him to slip while crossing a creek; lightning had struck him; his heart had attacked him—any of these could prove fatal to a man alone and exposed. They all feared that Randy was injured and unable to call for help because either he was incapacitated or he was in a radio dead zone, or his radio simply wasn't working. If that injury had occurred on the first day of his patrol, he would have been out there now for four days.

Lyness was only slightly perturbed that it had taken four days for them to gather. "Response time was *always* slow," she says, "largely, probably, because nothing ever happened to [the backcountry rangers] and because as of late, radios and repeaters had been unreliable."

Rather, she was floored by the fact that "someone actually followed some kind of protocol and *did* something. That," she says, "had not been the case in previous years."

Both Lyness and Durkee knew that Randy had been incommunicado for eight days just the season before while stationed at LeConte Canyon. "Can you believe that?" says Durkee, who had read the logbook in which Randy had penned his frustrations. On the sixth day without contact, he'd written, "How long before they come to look? There's a policy. . . ." After eight days: "Do I have a safety net? 8 days and counting."

Communication into the far reaches of the parks had always been an issue. In the 1920s and 1930s, hundreds of miles of telephone wire had been strung across the backcountry. Rangers at that time were trained linemen. If they needed assistance or spotted a forest fire, the standard operating procedure was to climb the nearest tree where the wires ran, tap in and hand-crank a message to headquarters. In most cases it would take days to reach outlying areas, a reasonable response time for that era.

A letter from John R. White, the superintendent of Sequoia National Park, to Colonel C. G. Thomson, the superintendent of Yosemite National Park, dated May 7, 1930, had discussed a new era in technology "with regard to the possibility of radio communication with outpost stations. . . . I learned that the Signal Corps outfits would easily communicate from any part of the Sequoia National Park to Ash Mountain headquarters. . . . However, I learned that the outfits . . . weigh approximately 650 pounds with the batteries [and] . . . it is too heavy for our purposes."

A few years later, when wireless radio lost a little weight, the idea of disassembling hundreds of miles of fairly reliable wire was met with resistance. On July 26, 1934, L. F. Cook, the associate forester of the National Park Service, sent the chief forester in Washington, D.C., a letter regarding the approval of 30 miles of telephone line between the Kern River ranger station and Crabtree Meadow in Sequoia. Acknowledging the difficulties of such an extensive line, Cook wrote, "I believe that field communication is very much needed for protection" of the natural resources and citizens using the park.

Regarding radio usage, he wrote, "I am not at all sold on its use as yet. . . . It appears to me that this form of communication is still very much in the experimental stage and a matter for experts to work with. Communications is so dependent on weather conditions, expert handling of equipment, on distances, and so subject to the little known complications that I would personally hate to have to depend upon this form of communication . . . as yet I have not seen a test made which was entirely dependable."

Sequoia Superintendent White echoed Cook's sentiment when he wrote the National Park Service's chief engineer on August 23, 1934. "My dear Mr. Kittredge: I have your letter of August 16 about the installation of the radio, and thank you for pressing this matter for us. I am, however, going ahead under the authority given me by the Director, in the matter of construction of a Kern Canyon telephone line. . . ."

I feel that no matter how much we perfect radio, it can never entirely replace the telephone line communication."

By July 1935, Sequoia's first radio set was installed at the Ash Mountain headquarters. On August 30, Superintendent White again wrote the chief engineer, no doubt with a sense of I told you so. "Dear Sir: As advised by my telegram, our radio headquarters set is completely out of commission. . . ."

Sixty years later the telephone lines had long been removed and radio technology still hadn't been perfected, and the thought that Randy might be out there in need of assistance and unable to call for help angered Lyness.

"The fact is, the whole radio thing was massively screwed up . . . and had been deteriorating for some time," she says. "Repeaters didn't work, radios didn't work—I had at least three radios that summer—so it was not unusual to not be able to contact someone. It just seemed not to be a priority for anyone who had the power to do something about it to get radio communication in order."

One of Randy's more cynical jokes struck a little too close to home that evening at Bench Lake: "If you're going to get hurt in the park, make sure you do it in a place where there's good radio coverage."

Ironically, in his 1995 end-of-season report, Randy had reiterated what he'd been saying for years: "Radio communication . . . was difficult again this season; everyone knows.

"We hope it'll be better next year."

A SOFT, LIGHT BLUE SKY held a few drifting cirrus clouds, wispy, elongated remnants from the afternoon thunderstorms. Soon the clouds would catch the setting sun's fiery reds and oranges that would bathe the basin's surrounding peaks in the glorious light for which these mountains are famous.

Normally the rangers welcomed the evening light, even planned their days so they'd be positioned, come sunset, in front of a monolithic hunk of granite or west-facing cirque—a backcountry hike-in theater. But come dusk on the day that Randy's SAR was initiated, there was no pleasant anticipation. The evening light served only to usher in the darkness that punctuated the end of Randy's fourth day without contact and another cold night for him. Alone.

Upon their arrival at the Bench Lake station, Coffman had instructed the rangers to read Randy's logbook to glean any information that might give them an idea where he had gone. As they huddled around the journal at the picnic table, they noted the places he'd already patrolled and conveyed them to Coffman, who was keeping a list of clues. Intermittently, Coffman threw questions into the mix: How many miles would Randy travel in a day while on a trail? While off-trail? Did he prefer to camp in protected, wooded areas or in the open? Would he scramble up and over a difficult class 3 ridgeline or take the longer but easier route around such a feature? The queries were indirectly keying the rangers into Randy's profile as a wilderness traveler—a psychology that would help them make more educated guesses as to his actions. Coffman encouraged ideas. "If you remember Randy mentioning someplace he wanted to check out during training, some peak he wanted to climb," said Coffman, "speak up."

During the course of the discussion, Coffman maintained radio contact with Dave Ashe back at headquarters and "inked up" the topographic map on the picnic table, dividing it into sixteen segments labeled A through P. Each segment was delineated by obvious geographic boundaries such as rivers, ridgelines, trails, meadows, and mountain peaks or passes. They were all within an area that was roughly 80 square miles, the area that the rangers agreed represented the outer limits of where Randy might have traveled on a four-day patrol.

This was when a lone backpacker strolled up to the station. The helicopter had just lifted off, and he greeted the rangers with a poorly timed "What's all the racket?"

Coffman approached the backpacker.

"I remember Coffman shot back something like 'Sorry about the noise, but we've got a missing ranger that could be in trouble,' " recounts Durkee. "I think he was trying to keep him from interrupting our focus while we were planning, but the guy was completely clueless and took off his pack like he wanted to visit, and started asking Coffman all these questions about fishing spots and rattlesnakes."

That was when Durkee stood, with the intent to rescue Coffman, but when he got to the hiker, "I lost it, just a little bit," says Durkee. Uncharacteristically lowering his voice an octave, Durkee said to the backpacker, "Maybe you didn't hear him. We've got an E-MER-GEN-CY here." With that, he turned his back.

"Sorry," the backpacker said, and returned to the trail.

Durkee returned to the picnic table and stared at the map. The sheer size of the search area sank in. "Oh, shit," was Durkee's reaction. "We're going to need a lot of help," said Graban. Coffman said, "It's coming." All agreed they were up against a daunting task. Search areas this massive were most often reserved for downed aircraft. A missing person on foot was usually much more limited in terms of mileage.

And on the map, the shape of the search area was anything but a nice circle or square grid spreading out from the red X that marked the "victim's" last known whereabouts. Such computerized representations are unrealistic in mountainous terrain. This search area's boundary lines were chaotic, like the terrain itself. The lines came together ungracefully and represented, at best, an incongruous shape that could have been drawn by a 4-year-old.

But just as a 4-year-old can see a rhinoceros or dinosaur through a scrawled assortment of lines, the rangers saw topographic familiarity beneath the ink. Erratic curves and squiggles represented ridgelines and cirques, elevation gains and losses; sweeping strokes were canyons carved by water; amoeba-like shapes were basins; the corridor of Cartridge Creek jutted away from the search area like an arm; the Muro Blanco, a boomerang-shaped leg, dangled to the south. But it wasn't the configuration of the search area that worried them—it was the sheer magnitude combined with the ruggedness of the terrain. A geographic monster pieced together by hazards that could swallow a man.

It wouldn't be the first time.

Entire airplanes and their crews had crashed in the High Sierra and were still missing. Others had taken decades to be found. During two weeks of December 1943, four B-24 Liberator bombers from the 461st Bombardment Group on accelerated training for deployment to the European theater of World War II had crashed in winter storms. A massive air search had been launched, but not a scrap of wreckage was found. One of the bombers, which had last been reported between Las Vegas, Nevada, and the eastern Sierra foothill town of Independence, became a legend to Sequoia and Kings Canyon backcountry rangers in ensuing years because of the father of its 24-year-old copilot, Second Lieutenant Robert M. Hester.

Clinton Hester, convinced that the Liberator had crashed somewhere in Kings Canyon, was determined to honor his son's service by bringing his body home. Year after year, he searched the high country himself, with any volunteers he could recruit. His quest was relentless, but after more than a decade of searching, he had not discovered a single clue to substantiate his theory.

In 1959, after fourteen years of methodically combing the mountains, Clinton Hester died from a heart ailment. One year after his death, in July 1960, his son's bomber was found by a ranger on a geological survey in the Black Divide range near LeConte Canyon. The plane had crashed into a 12,500-foot peak and exploded on impact. Some of the debris ended up in what would be known as Hester Lake. The elder Hester had come within a few miles of the wreckage.

The Hester Liberator was often referenced to illustrate just how overwhelming a search for a person in the High Sierra could be, considering that the wreckage from a 70-foot-long shiny silver bomber had eluded detection for a decade and a half.

Still, there was only one story in the parks' history of a missing person on foot who had not been found during the course of a search-and-rescue operation. His name was Fred Gist, a 66-year-old real estate appraiser from San Luis Obispo who had disappeared just beyond the southwest boundary of the Bench Lake ranger's patrol area on the Monarch Divide. Gist was last seen by his companions on August 19, 1975, near where Dougherty Creek flows into the crystalline waters of the Lake of the Fallen Moon.

Five search dogs and four trackers from the U.S. Border Patrol joined twenty-six rangers and volunteers from across the state on an intense, leave-no-stone-unturned search that began two days after Gist's disappearance. It was learned that Gist had been packed in with horses and wasn't a particularly strong hiker, so the search area was fairly compact, about three miles across. Using classic strategy of the time, it was grid-searched with dogs; according to the case incident report, "not a trace of the missing person was found."

The search was called off on the seventh day, after high-resolution military photographs of the area produced no results. Fred Gist's fate had been a mystery for more than a decade until backpackers found his skull near Dougherty Creek. Without any knowledge of the mystery, they left the skull on the doorstep of the Simpson Meadow ranger station with a hand-drawn map showing where it had been found. Ironically, they had named the skull "Fred."

The Gist search illustrates, perhaps more so than the Hester plane crash, how these mountains can hide a person, even with trained search teams combing an area. There was, however, one significant difference between Gist and Randy. Gist was wholly unprepared for the freezing nights; he reportedly didn't even have a sleeping bag with him.

Not only was Randy extremely fit, he also had with him survival gear and the knowledge to use it. He just had to hang on and let searchers find him somewhere in the 80-square-mile search area. No doubt about it, he was the classic needle in a massive haystack—but times had changed and so had modern-day search techniques.

In 1976, shortly after the Gist search, a lieutenant colonel in the U.S. Air Force named Robert Mattson came up with a brand-new method for prioritizing ground search areas. His innovative strategy, first published in the spring 1976 issue of *Search and Rescue Magazine,* came to be known as the Mattson method or the Mattson consensus. It was inspired by the pioneering work of B. O. Koopman, who, as a member of the U.S. Navy's Operations Evaluation Group, created a mathematical approach to locating enemy submarines in the vast oceans during World War II. So effective was the strategy, Koopman and his group were credited with being key to winning the battle against German U-boats in the Atlantic.

The Mattson consensus has, for the most part, remained the favorite strategy of SAR professionals, including Coffman, who implemented its classic approach as the leader of the search effort at Bench Lake.

According to Mattson, experts who knew something about either the missing person or the terrain should be brought together; these individuals should be "the most informed and experienced personnel available." In this case the rangers knew both Randy and the Sierra. After collecting as much information as possible about the victim and the area, Coffman divided the overall search area into reasonably sized segments. Then, using a secret ballot, each ranger assigned each segment a number value—high for areas where Randy most probably was, low for least-probable areas. According to Mattson, it was "best to do this privately because it will insure [sic] that even the meeker individuals will be able to express their opinion without being intimidated by the more vocal members of the group."

Though Coffman ran the show and knew the history behind the theory, the rangers knew the drill and spoke the same acronym-heavy language. POA, for example, was "probability of area," the probability that Randy was in a certain segment. ROW stood for the "rest of the world" and considered the possibility that Randy was somewhere other than inside the designated search area.

The percentage points assigned by each ranger for sixteen segments plus the theoretical ROW segment had to add up to 100 points. Nobody could assign a zero for any segment. That would mean they knew with certainty that Randy was not in that particular segment, which was impossible. In his article, Mattson had taunted readers for such optimism in the face of unknowns: "If you KNOW where the survivors are, why are you searching!!!!!???!!!!"

The overall message Mattson conveyed, above and beyond the mathematical approach, was "Never discard information, keep an open mind, use common sense, and dig, dig, dig for information."

And dig they had. Coffman had a notebook full of notes to prove it.

In his logbook Randy had reported going south on the John Muir Trail to Pinchot Pass twice, once to the summit and the second time over the top to Woods Creek. Acting on their knowledge of Randy's habits as a ranger, they deduced that it was unlikely he'd gone that direction again—either via the JMT or any cross-country routes that eventually met up with it in that southerly direction.

On the other hand, Randy had not yet been to Lake Basin—which Durkee, Lyness, and Graban knew was a sacred place for him. Nor had he covered the cross-country routes in Upper Basin or any of the tucked-away gems north of the Bench Lake Trail, including Dumbbell Lakes and Marion Lake. Using this line of reasoning, the rangers threw out ideas of probable distances and places Randy might have visited on a three- to four-day patrol.

The information-gathering process had taken hours, but the voting process took about twenty minutes. Not surprisingly, the Lake Basin area (Segment F) was unanimously valued as the highest-percentage POA at 26.20 percent, while Marion Lake and its surrounding cirque (Segment G) was the second-most-probable consensus at 19.20 percent. The ROW option was voted as the lowest POA for everybody except Durkee, who assigned that option a curiously high percentage compared with the other rangers. The higher value prompted Coffman to ask, "You think Randy might have left the park? Why?"

"I told Coffman that Randy's life was in turmoil," says Durkee, "though I didn't go into details with Lo sitting right there next to me." Durkee also kept quiet about what he described as a "very slight, but unshakable" suspicion that his friend might have gone off to some special place and ended his life.

After Coffman dismissed them till morning and the other rangers had wandered off to their respective sleeping spots, Durkee made a discreet detour to the door of the station. Randy's note was still pinned to the canvas. The date he'd written was June 21. But it was July. Everyone else had discounted the mixup of "J" months as an honest slip of the pen, but Durkee couldn't stop thinking that it was a potential clue to Randy's mindset at the time. He reprimanded himself for his para-

noia and pushed aside the tent flap. As always, Randy's residence was
spartan. "Randy never was much for putting up pictures or drapes to
make his stations more homey," says Durkee. "It was a minimalist base
camp."

Quickly, Durkee's headlamp beam found its mark: the steel footlocker where he knew Randy would have kept his sidearm. As expected, it was padlocked. He gave the lock a tug, just in case. Solid.
He then turned his attention to Randy's military field desk—an olive
drab rectangular wooden box with a leather handle on either end. The
front was a row of drawers and cubbyholes topped by a worn-smooth
working surface that folded up to reveal a storage compartment. Inside
he found the expected stacks of mandatory reading: the new NPS 9
Law Enforcement Policy and Guidelines binder; a few inches' worth of
backcountry policy, which Randy could recite from memory, mostly
because nothing had changed much in the past decade; some EMT
refresher manuals; a stack of citations; and the recently proposed, but
not implemented, meadow management plan that had been dispersed
during training to some of the backcountry rangers. Atop this particular document was a pen, and within the pages were Randy's notes and
suggestions.

"It was a work in progress," says Durkee, "which told me Randy
intended to come back." With that rationalization, the mentally exhausted ranger retired to his tent.

Coffman continued to plan into the night. Dave Ashe was his point
man in the frontcountry, to whom he relayed—among other things—
the results of the consensus.

Ashe and another ranger named Scott Wanek had organized an impromptu incident command post at the Kings Canyon fire station. They
transformed a dormitory into a planning room and began the process
of spreading the word to various state emergency response groups—a
network of organizations that included the California Rescue Dog Association (CARDA) and volunteer SAR teams from different counties
throughout the state. The military and state highway patrol were put
on standby, with potential requests for air support and personnel. The

emphasis in requesting personnel was on expert hiker skills. Coffman had told Ashe to make it very clear: "The search area is complicated, dangerous, off-trail terrain." Ashe, in turn, conveyed that he wanted "quality, not quantity." The underlying message was "We don't want to have to rescue the rescuers."

Somehow, even with all his other duties, Ashe found a few minutes to prep CASIE for data. CASIE, or Computer-Aided Search Information Exchange, was a program designed to simplify most of the calculations related to managing a search emergency using modern search theory and terminology. Once it has been plugged into CASIE, an overwhelming search area and operation becomes more easily digestible by date or segment. A glance at a computer printout provides basic information about the manner of searching a certain segment (air, foot, dog team, etc.) and how effective the searchers believe they were in "clearing" that area. Using this method to keep track of a large mass of land, the leader of the search—in this case, Coffman—would cross off search segments once he felt confident they were clear.

This, of course, presumes the missing person is not on the move and has not reentered an area already cleared, the reason for the "hug a tree" strategy preached at wilderness survival classes. Another difficulty is that segments are generally considered surface areas—not underwater, underground, or under a rock slide. In an area as vast as the Morgenson SAR, where only two segments were smaller than 500 acres, the majority were around 2,000 acres, and one segment was initially more than 7,000 acres, a thorough surface search was difficult enough. Compounding the challenge, the high country has a myriad of streams and rivers that empty into hundreds, if not thousands, of lakes. Nearly every peak has dozens of active rock-slide and snow avalanche paths, any of which could bury or otherwise conceal an injured or deceased victim.

Randy could be a few yards from a shouting search team, yet not be discovered. The same team could be employing an air-scent-trained search dog but if Randy was downwind, the dog would not catch his scent. A search-dog handler described the nature of a SAR in the High

Sierra as an "organized search in chaotic terrain." That description didn't even begin to explore the depth of the chaos if one other possibility was included.

What if Randy didn't want to be found?

CHIEF RANGER DEBBIE BIRD was saddle-sore and weary when she arrived at the Road's End trailhead just after dark. The horseback ride from Vidette Meadow, where she'd last seen ranger Lo Lyness, was about 16 miles. She'd seen the helicopter activity and even without a working radio had deduced that a search-and-rescue operation for Randy had been initiated.

She drove immediately to Cedar Grove, where she saw the fire station lit up and alive with activity. She looked for Coffman, figuring he'd be in charge of the search, but instead found Ashe, who brought her up to speed on the few details currently available. "Coffman," he told her, "is at Bench Lake."

This wasn't what Bird wanted to hear. She felt that Coffman and Ashe weren't thinking far enough ahead—not treating this as a major incident. She sensed the search would "evolve into something more," she explains. "Once these things get going, it becomes very difficult to catch up logistically." As the chief ranger, Bird could pull rank and issue an order to use an incident command system (ICS), but with Randy Coffman, "who is a, well . . . a very talented SAR ranger," she says, "it was much more effective to work at making *your* idea become *his* idea, rather than issuing some kind of order."

With this in mind, Bird radioed Coffman, whom she hoped would be receptive to an ICS.

The incident command system was developed in the 1970s in response to a series of fires in Southern California during which a number of municipal, state, federal, and county fire authorities collaborated to form FIRESCOPE (Firefighting Resources of California Organized for Potential Emergencies). During that firestorm, a lack of coordination and cooperation between various agencies had resulted in overlapping efforts and, worse, major gaps in the response. Consequently, some areas were overstaffed with firefighters, while under-manned property nearby was destroyed. From this experience came the original ICS model for managing wildland fires and emergencies at the city, county, and, eventually, federal level. In 1985, the Park Service adopted and pioneered the ICS for search-and-rescue and other emergency operations and is widely credited with honing the system.

Whenever the evening news reports that a search-and-rescue operation is in progress for a missing person, in wilderness terrain or otherwise, an ICS is almost certainly the skeleton that is moving the body of the search.

Bird was concerned that Coffman was initially performing the duties of incident commander, operations chief, and planning chief, which was "way too much for one person to handle effectively" for any length of time.

"I wanted Randy [Coffman] to stay in the frontcountry and plan the search, not go into the backcountry himself and be part of the search on the ground," says Bird, who explained it like a military battle: "The general doesn't go out with the troops and lead the charge. Instead, it's his job to stay a ways back and plan the execution of the battle plan, including factoring in events as they change. This means collecting intelligence, ordering the tanks and aircraft that are needed to execute the plan, and making sure you have enough troops to do the job and have a way to transport, feed, and house the soldiers."

The chief ranger wanted Coffman to either assign the incident command position to someone else or "come out of the backcountry and start thinking about planning the full-scale search," which included delegating duties to qualified personnel.

When Bird heard Coffman's voice over a static-filled radio connection, he was receptive to the idea of an ICS but had no interest in relinquishing his command of the SAR. At this point, it was personal—he felt a sense of responsibility to see this one through. Incident commanders often wear numerous hats initially in a SAR, and for good reason. The ICS was formulated around the basic principle of accountability. In the end, the weight of the operation's success or failure was on his shoulders.

Confident that the search was on track, Bird called Randy's wife, Judi Morgenson, with whom she had worked as a wine steward at the Ahwahnee Hotel in Yosemite during the early 1970s. Bird, who had begun her career with the NPS well after Randy's first summer seasons, was proof positive that a high-level permanent position was attainable in the Park Service if you had the skills, the desire, and a willingness to jump through hundreds of hoops. Here she was in 1996, filling the boots of the once male-dominated chief ranger position at one of the country's wildest national parks. In the past, she'd dealt with a lot of unsavory duties in her rise up the duty chain, from arresting drunks to being bitten, stepped on, and kicked by mules, to recovering bodies to shooting human-attacking coyotes to dealing with the mountains of paperwork in constant upheaval on her desk.

But it was a phone call like this, regarding the uncertain well-being of a ranger who ultimately was under her command, that proved to be one of the most difficult duties she had performed in her entire career.

RANDY HAD PACKED his gear for the 1996 season with the belief that Judi would not be welcoming him with open arms at their Sedona, Arizona, home come October. He loaded his Toyota truck with more boxes than usual, intending to store favorite books from his parents' library, his camera equipment, extra clothes, ski equipment, and photo albums in Bishop, California, on the eastern side of the Sierra.

He had always given Judi books as gifts, often before heading into the backcountry. Generally, they held deeper meaning than just a good read—a message to Judi, something he felt strongly about and wanted to share. In seasons past, they had provided her a measure of comfort on the lonely nights spent at home while he was away for months at a time. "It's hard to explain, but reading a book Randy left me was kind of like having him there next to me," she says. "After so many years, we were in tune with each other. Books that spoke to him usually spoke to me too."

Even with divorce papers stashed in his backpack, Randy continued this tradition. He had given Judi *I Heard the Owl Call My Name,* a novel by Margaret Craven. The gift had been an unexpected and tender message, despite the pending terminus of their marriage. It reminded her, if only for an instant, how charming and sweet her partner in life could be. But it was a short-lived respite. After Randy drove off, she placed the book on her nightstand, where it sat unread.

Almost two months later, on the evening of July 24, she still hadn't cracked the cover when the phone rang and a familiar voice greeted her.

"Judi," said the chief ranger, "this is Debbie Bird at the park. You haven't heard from Randy recently, have you?"

The last time Judi had heard from Randy was during ranger training. "Not since late June," she responded. In that phone conversation, Randy had pleaded with Judi to join him in the backcountry. He wanted to make things right—if she would have him. But Judi had held her ground, letting Randy know that it wasn't that simple.

Bird informed Judi that Randy was overdue from a patrol, and a search was in progress. "He could just be having radio problems," said Bird, "but it's been four days."

"Four days?" said Judi. "He's trying to worry me."

Judi then told Bird that they had separated and she had filed for divorce. This was news to Bird, and she became concerned that perhaps Randy had hiked out of the mountains. If that were the case, the well-being of the searchers was unnecessarily at stake.

"Is there anyplace he'd go? Anybody he'd call?" asked Bird.

"Alden Nash," said Judi. "Or maybe our friend Stuart Scofield."

"We can contact Alden," said Bird. "You might want to follow up with your friend and let us know if he's heard anything."

Bird gave Judi her direct lines at headquarters and at home and told her she'd call with an update the following day.

Judi ended the conversation by telling Bird, "It wouldn't be like Randy to leave the mountains. That's where he needs to be right now."

Judi wasn't overly concerned. On the contrary, she was perturbed. She became fixated on the suspicion that he was merely taking his time

on the patrol, lollygagging around, looking at wildflowers and purposely ignoring the radio. His motive? To get even with her for not
coming into the mountains. For not giving him another chance.

"Randy would have known I'd be the first call they'd make when he
was overdue," says Judi. But he was mistaken if he thought this stunt
was going to make her lose any sleep. Picking up the address book
she and Randy had filled over the years, Judi searched for Scofield's
number.

In the living room of their home, she sat in her reading chair and
reached for the phone. As if on cue in a campy horror flick, thunder
rumbled and lightning flashed, lighting up the desert. Randy always
loved storms—in the desert and the Sierra—and for a moment she was
taken back to another storm they experienced early in their marriage.

They'd been hustling north on the John Muir Trail, rushing to make
it back to Randy's Tyndall Creek cabin before the storm was upon
them. To the west across Kern Canyon lightning strikes were hitting
the jagged teeth of the Kaweah Peaks as menacing dark clouds closed
in overhead, growling and illuminated with pulses of electricity. When
they moved quickly through a sparse scattering of foxtail pines, one
exploded with a deafening crash. The trees thinned as they climbed,
and Judi grew terrified when they emerged onto the treeless 11,300-
foot Bighorn Plateau. They squinted in the howling wind, watching
snow squalls blow across the tundra. The safety of a distant treeline
and a lower elevation was more than two and a half miles across completely exposed country. Judi hiked as fast as her legs could carry her,
wanting to pass Randy, who strolled casually ahead, enraptured by the
tempest.

The rain and snow turned to hail, which pelted Judi and stung her
face. Frozen granules accumulated in the indentation of the trail, ren-
dering it a white line to safety that she was fixated on.

But a quarter mile out on the plateau, Randy's aura soothed her and
she was able to relax into the experience. Even though lightning was
striking the plateau and the hair on her neck stood on end from the
static, she suddenly felt safe. Judi observed that the world was coming

down around Randy, yet he was calm and composed in the face of the storm, not unlike a battlefield hero seemingly immune to flying bullets. They made it to Tyndall Creek wet, exhilarated, and happily unscathed—an initiation for Judi, but for Randy just another walk in the park.

Phone in hand, Judi dialed Scofield, repeating in her mind what she'd told herself every summer for more than two decades:

"Randy can take care of himself in the mountains."

CHAPTER FIVE

AFTER THE RIOTS

But ever since I was old enough to be cynical I have been visiting national parks, and they are a cure for cynicism, an exhilarating rest from the competitive avarice we call the American Way. . . . Absolutely American, absolutely democratic, they reflect us at our best rather than our worst.

—*Wallace Stegner, 1983*

Randy could make a swarm of mosquitoes seem like the most romantic thing in the world.

—*Judi Morgenson, 2002*

AT THE END OF THE 1971 summer season, Randy came down from the mountains tanned, lonely, underweight, and hungry for his mother's cooking.

Esther Morgenson was managing The Art Place, one of Yosemite's art galleries, which also sold homemade cards, candles, and other crafts. Randy wandered in looking for her, but was sidetracked by the resident candlemaker, Judi Douglas, a self-proclaimed "city girl" from Orange County.

"So, what brought you to Yosemite?" asked Randy after he had introduced himself. Judi, instantly smitten by the dashing young ranger, explained how she'd taken a year off from studying art history at San Jose State University to travel, party, and see the galleries and cathedrals of Europe with her best friend, Gail Ritchie. Then, with a couple of months left before the next semester, she'd decided to follow Gail to Yosemite, where Gail had helped Judi get this job at the gallery. Randy nodded his approval and told her she'd been wise to leave the best for the end of her journey. The look he gave her conveyed that he meant something more than Yosemite. "Yeah," says Judi, "he charmed the socks right off of me from the start."

Judi had gotten to know the tourists' Yosemite, but Randy took her to hideaway beaches on the Merced River that weren't on tourist maps; they also had drinks in the rustically romantic Ahwahnee Hotel and shared stories of their travels. "Randy could paint pictures with words," says Judi. "He'd take me on trips to India, Nepal, and Japan in one conversation."

A little more than a week after they'd met, Randy rendezvoused with Judi and Gail at the art gallery to view an exhibition of Asian artwork that made for another night of easy conversation. This was Judi's domain, and Randy listened intently as she explained some of the history behind the pieces. "He made me feel important," says Judi. "That what I wanted to do with art was important—not just special, or neat, but important. And he never made me feel rushed, which wasn't easy to find back then. Most guys couldn't wait to get out of an art gallery. Randy just strolled through soaking in the art.

"First, we were inspired by this amazing artwork, and then we walked outside only to find ourselves in this amazing natural amphitheater." The granite cliffs glowed in the moonlight, the pine trees shot up into a sky full of stars, and Judi and Gail had Randy between them, the three happily holding hands as they headed toward Judi's dorm room at Happy Isles.

"Gail let go," says Judi, "and I never did."

It wasn't long before Judi was a frequent guest at the Morgensons' residence on the Ahwahnee Meadow, drinking cocktails in the yard and watching evening light on Half Dome before being called in to dinner. The protocol of the Morgenson household was a bit surreal for Judi, whose own family life was more unstructured. She was impressed by Esther's attention to detail—how she wrapped everything she put in the freezer slowly and artistically in plastic, like Christmas gifts. At the dinner table, Judi appreciated to the point of awe the image of Dana at one end carving the meat and Esther at the other end serving the salad, of plates passed from one person to the next until everyone was served. Then, and only then, did anyone lift a fork.

Judi learned quickly that Esther was the sensitive and quiet artist while Dana was the engaging diplomat with a confident command of language. Randy, she observed, had inherited both his mother's sensitivity and his father's way with words; he was also handsome, with thick, dark hair and expressive, gentle eyes that held a hint of something she couldn't put her finger on. Mystery, perhaps. By the time her job ended and she had to return to the city for school, Randy had captured her heart.

HIS OWN HEART FILLED with romance, Randy strove to get his photography and his stories published. He'd been taking notes in the backcountry and made dozens of prints from both his travels abroad and the Sierra that he sent off to various magazines. He read the letters he'd written home, all of which his mother had typed up, and reviewed his journals. And what more inspiring setting to write one's memoirs than staring off toward Half Dome as the first winter snows dusted the valley?

As temperatures dropped and tourists fled, an unlikely mentor arrived—perhaps to research a book or to visit Ansel Adams, or likely he just knew that this was the best time to come to Yosemite. Perhaps he came to see Dana Morgenson, who had become a bit of a celebrity himself in botanical circles. Regardless, Wallace Stegner ended up in the Morgensons' living room that autumn of 1971.

He'd just won the Pulitzer Prize for his novel *Angle of Repose* and was about to begin his final year teaching at Stanford University, where he'd founded the creative writing program in 1946. In the ensuing decades, he had taught hundreds of students, some of whom had been awarded a Stegner Fellowship: Wendell Berry, Ken Kesey, Ernest Gaines, Raymond Carver, and Edward Abbey.

During the course of eco-fired conversation—snowmobile usage in the national parks and other such controversies of the times—Stegner learned of Randy's aspirations as a writer and offered to read something he had written. Randy had just refined two stories, "Little Town of Golapangri" and "Within Mountains," both of which left Yosemite Valley in Wallace Stegner's satchel.

By Christmas, Randy had received a handful of rejection letters for other articles he'd submitted to magazines, but he had heard nothing from Stegner. He'd almost given up when, after the new year, it arrived: two single-spaced typed pages on the letterhead of Stanford University.

"I've been interested to read your two pieces," wrote Stegner on January 26, 1972. "They're literate, sensitive, and earnest, and they probably made you feel good when you wrote them. . . . But, I doubt they'll do for publication. . . . I'll try to tell you why, always with the risk of being blunt and seeming unfriendly."

First Stegner commented on "Within Mountains," which had been inspired by LeConte Canyon:

It's all about your feelings and sensations, and those don't communicate well to a reader. He half feels you indulging a sort of yeasty nature mysticism, and he has no way to join you in it because you give him nothing concrete enough to see or smell or touch or hear. Your pictures are all generalized. . . . There is no foreground, middle ground, background; there are not even details, but only generalizations of details. 'As warm sunlight reaches into the canyon, chipmunks and chickarees chase over Sierra slickrock and across dried grassy places.' . . . It is a generalized sunlight hitting not a place . . . but a generalized Nature crossed by anonymous and generalized little beasties. I would feel the emotions this . . . arouses in you if I could put myself in your shoes. . . .

And so through the whole essay. You come at us emotion-

first. You try to evoke the emotions without ever giving us the particular place, picture, actions, sense impressions, that might let us feel as you do. Does this mean anything to you? It may be that you're feeling the influences of Muir, who did get away with a lot of generalizing of that kind. But he got away with it partly because he had a sort of exclamatory genius, he was a whirling dervish of Nature, and not all of us can be anything like that. . . . If I were you, I'd follow Thoreau or somebody like that rather than Muir. . . . You may take it as my version of an almost-infallible rule of thumb that nature description by itself is very hard to get away with—it's necessarily pretty inert and undramatic. . . . You know a lot about the mountains that doesn't show here. If you tried to tell us what you know, your feelings about the mountains would probably come across more strongly than they do when you work on the feelings exclusively.

Stegner then gave an equally honest, scathing critique of "Little Town of Golapangri." He ended the letter by reiterating:

I do think that you have to steer away from the general, quit looking at the heavens and thinking large vague thoughts and feeling large vague feelings, and start watching the pebbles and ants and sunshine and shadow right under your feet. If you can learn that, and discipline yourself to keep remembering it, you can do with words what you obviously want to do with them.

Good luck. My best to your family, who were one of the pleasantest things about Yosemite. And just a word: A French friend who yesterday blew in from Paris tells me that France has totally banned snowmobiles, anywhere. Who are we that we should trail behind the French? What's the status on snowmobiles in the park, now? Any decisions?

Best,
Wallace Stegner

Randy replied promptly.

Dear Mr. Stegner—

*I owe you a large thank-you for . . . the honest worth of your
criticism. Both barrels is what I've been asking for from many
quarters for some time, without getting it. I've no illusions about
literary genius, though I remain confident enough in a rapport
with words to hope I could reach some success in writing with a
little helpful guidance. Thank you.*

*I would like to send you something again when I get it to a
point that seems right, which may be awhile, if I may . . .*

In closing, Randy continued the snowmobile banter:

*Apparently there is no change in status of snowmobiles here.
They are permitted on the Tioga and Glacier Point roads only;
the latter is patrolled by a skiing ranger, the former is not . . .
perhaps in this age of motors we should be happy they are as
restricted as they are, if the restriction works. I've not talked
to anyone who knows that snowmobiles are roaring through
forbidden territory in the park, but even so I can't let go of the
dozen things wrong with any use of them here.*

Within a month, Stegner replied:

Dear Randy,

*I'm glad you weren't permanently disabled by my criticisms,
and that you're going on to try other things and other ways. The
literary business is a contact sport—you have to like bruises and
knocks to stay in it.*

These letters were the beginning of a string of correspondence that would span years. Stegner continued to encourage Randy to send him his writings, while Randy kept Wallace abreast of issues within the National Park Service, fishing around for information and becoming a sort of stringer for Stegner's kindred bent toward environmentalism.

On March 2, 1972, Stegner wrote:

Dear Randy:

Many thanks for the latest dope on snowmobiles. The NPS may mutter about my misunderstanding, hearing "park" when they said, "Valley," but it ain't so. There was an effort, I'm sure, to shut the machines out of the whole park, and that's what the director mentioned to me. But I guess the lobby was too strong, or fear of it was. In any case, Morton's recent announcement about limiting visitation in the wilderness areas of Great Smokes, Sequoia Kings Canyon, and Rocky Mountain is a hopeful sign for the future. So is the really drastic shutting-down on the poisoning of wildlife on federal lands. Stan Cain, who used to be Assistant Secretary under Udall, and was unable to stop the Fish and Wildlife boys from their poison baiting, was here the other night for dinner, and is very optimistic that the battle is nearly won. So cheers. It'll only take four more lifetimes. Meantime the desert will be all plowed under by off-trail bikes, long before the high country is gone. Retreat upward.

Best,
Wallace Stegner

BY THE SUMMER OF 1973 the only thing Randy had "published" was this handwritten note he stuck to the door of his ranger station at McClure Meadow:

Welcome to John Muir's Range of Light, to Kings Canyon National Park.

Only one tiny request—please respect and care for these
mountains. Especially refrain from discarding litter. Foil, cans,
and glass will not self-destruct. They remain for years. Please
don't try to burn foil in your campfire. No one likes to see garbage
in the mountains, least of all you. And we have no garbage
collection service here. Only we can preserve these mountains in
a natural and beautiful condition for us to continue enjoying.
It will be a very long time before an Ice Age cleans them of our
tracks, or new mountains are created in their place.
    Please, please consider enough the life of a pretty meadow
to refrain from camping in any, particularly from building a
fire on one. Would you build a fire on your own lawn? At these
high elevations a burned spot will probably never recover. There
will always be the ugly mark of your campfire. I ask you to leave
beauty as you walk through the mountains. HAVE A GOOD
DAY!

Randy Morgenson, Evolution Ranger

It wasn't Pulitzer material, but word got around. His journal en-
tries from Rae Lakes in 1965, Charlotte Lake in 1966, LeConte Canyon
in 1971, and McClure Meadow in 1972 were good reading, not just
the standard logbook fodder recorded by most of the rangers that included miles hiked, weather reports, number of people contacted, citations issued, medevacs performed, lost Boy Scouts found—oftentimes
written with that much detail, and no emotion.

"Then Randy came along and started something. He put to words
what we felt in the backcountry," says one of his fellow rangers.
"Nobody had really done that, but he took care to convey his emotions. You couldn't put down a logbook written by Randy without
knowing, full well, his love for these mountains—and all the crap he'd
put up with as a ranger." It was, in fact, Randy's suggestion not only
that these logbooks be archived at park headquarters but also that they
be photocopied and kept in files at the individual duty stations where

they'd been written, so that future rangers could reference them for a "ranger's point of view" of the country they'd be taking care of and the problems to anticipate.

Whereas Randy's writing career had yet to bud, his relationship with Judi had flowered. The season after they'd met, he'd asked her to "walk in" and see him at McClure Meadow. The walk in had been a rude awakening for Judi, who toughed it out through 18 miles of wilderness and was still 9 miles short come nightfall. The bats were out and the birds had gone quiet when she was "taken in" (according to Randy; "rescued," according to Judi) by another ranger whom Randy had asked to keep an eye out while she hiked through his territory.

The following day Judi could barely walk she was so sore, but she hit the trail early in order to impress Randy. He met her en route, making the remaining miles to McClure Meadow dissolve by telling animated stories of every plant, creature, and rock they passed. That night, the hike in was forgotten. He made her dinner while she soaked her feet in a pail of water. She slept in his arms, feeling as safe and content as she'd ever felt in her life.

IN OCTOBER 1973, Randy and Judi took their first vacation together, a three-month road trip focused in southern Utah, where their lovers' bond was strengthened in anything but comfort. There was no semblance of civilized courting or chivalrous romance of the box-of-chocolates sort. They camped in dry desert washes off the sides of rutted roads. Baths were often a bucket of cold water because Randy strove to avoid burning even dead wood, knowing he was taking away from the life-sustaining nourishment the desert soil desperately needed. They stayed off paved surfaces completely for almost a month, until Randy decided to share with Judi a special place he associated with his youth: the Four Corners Monument, where Colorado, New Mexico, Utah, and Arizona come together.

As they drove, Randy recounted the trip to Judi. He had been 11 years old, sitting with his brother in the backseat of his parents' 1940 Buick LaSalle. He had swallowed dust on that washboard of a road all day, bouncing and lurching until the road deteriorated into two sandy tracks—parallel lines through the prairie grasses that became so deeply rutted the LaSalle could go no farther. At that point, they walked through the desert until they reached a large cairn of stones, with four lines of rocks leading away a short distance, representing the state lines. Randy explained how he felt there was magic in that spot, and his father, like a child himself, ran in circles around the pile of rocks, calling out the states as he stepped in each one, "Utah, Arizona, New Mexico, Colorado, Utah, Arizona . . ." It was infectious; the rest of the family joined in and they all ended up out of breath, laughing, just the four of them in the desert.

More than twenty years later, Randy and Judi retraced his childhood trip on a wide blacktop road "so smooth and straight it's a fight to stay awake on a hot day," he wrote in his diary.

They camped well off the paved road on the remnants of what Randy surmised was the original road he'd driven with his parents. As the sky darkened and the ground turned white with frost, they noticed a dim yellow glow ahead. From the warmth of their sleeping bags in the back of the pickup truck, the two mused: Was this glow the Four Corners Monument become an "all-night interpretive display"? Or was it a heated "comfort station with flush toilets and camper-trailer hookups"?

In the morning, they headed out on Randy's second pilgrimage to the Four Corners, "to take a measurement of the progress the nation makes toward providing access to all its corners for all its people." What Randy found in the place of that once-magical cairn of rocks he described as "a crime against land and against people, a mockery of human intelligence and a true measure of our worth as a culture upon the hapless surface of our gentle planet.

"There is no Four Corners anymore. They've <u>Paved</u> it!

"As Glen Canyon has been supplanted by a lake, the Four Corners, the plot of ground where 4 surveyed states come together has been supplanted by asphalt and cement. The Four Corners has been paved and ringed by canopied picnic tables, trash barrels and pit privies, paper sacks, Kleenex, and cans. A wide cement platform covers the actual spot. Consider it a moment. In celebration of the only spot where 4 states' corners meet, we have poured upon it concrete and asphalt. That says more than I could ever write about America. Destroy it to celebrate it.

"But we missed a bet. As usual we bungled it. The spot should have been asphalted, or cemented at ground level. Then we would have provided the traveling, tax paying public the only spot in the U.S. where we could be in 4 states at once without even getting out of our car. Bumper stickers could have been sold, and proudly worn by those who accomplished it. It's sad the engineers have so little taste. Thrills for the common motorist are becoming increasingly difficult to collect."

Combine this soured memory with the "smurky" horizon they encountered while approaching the power plants of Farmington, and Randy was spewing equal parts piss and vinegar in his diary. "I'm sick. I want to vomit, and swear at the top of my lungs with all my vocal strength, or better, destroy something, like a dam, or a coal-burning power plant."

In his eco-vehemence, Randy was ahead of his time. It would be another two years before environmental activism (or eco-terrorism, depending on your perspective) would be brought into the mainstream literary limelight via the monkey-wrenching mayhem of Edward Abbey's loosely fictional heroes right there in southern Utah.

Judi knew early on in their relationship that she had fallen in love with a man whose heart would always belong to the wilds. He'd dropped plenty of hints. "If I can manage it, I'll be in the mountains every summer for the rest of my life," he'd told her, "and you'll come with me, won't you? You'll visit, won't you?" But there had been times when his comments were more like warnings: "You know the mountains are in my blood? They'll always call to me."

But Judi was fine with Randy's bond with the wilderness, because as much as she loved being together, she had an independent streak of her own that made their union complete. Solitude, she'd found, nurtured her artistic creativity. Just as he wandered the hills, she found galleries to be her temples and cathedrals. And now, thanks to Randy, she'd been introduced to the wilds. Her journeys with him in the High Sierra and the deserts of the Southwest would forever be a source of inspiration, resulting in ideas that she was confident would ultimately be on display in art galleries. She knew it just as surely as she knew she'd follow this man, this ranger with a confident disposition, anywhere.

RICK SMITH WAS one of the charter members of the Yosemite Mafia. Smith began his career as a seasonal ranger for eleven years in Yellowstone. Then, after two years as a Peace Corps volunteer in Paraguay, he was hired in Yosemite as a seasonal; a few years later, shortly after the riots, he became permanent. An expert skier, he was promoted to the position of Badger Pass ranger and Tuolumne district ranger beginning in the winter of 1974–1975, the same winter Randy applied for the position of Nordic ranger—one of the more physically grueling, adventurous, and coveted jobs in Yosemite. Two positions were available.

Smith was known for having a discerning eye for talent and, like his Mafia brethren, didn't settle for mediocrity in any of the positions he supervised. He was always on the lookout for applicants who were likely to continue on as permanent rangers. The standard question he asked himself was: "Is this person capable and of the right mindset to eventually rise to the occasion and take over my position? Is this person a lifer in the Park Service?" If the answer was yes, he'd pull the application from the stack.

In 1974, the stack was massive. There weren't many winter jobs available in Yosemite, so the summer seasonals who wanted to stick around scrambled for them, fluffing their résumés with all conceivable qualifications. At the time everybody wanted to be a ranger, and some jobs in Yosemite had more than 1,000 applicants. It was easy to find parking lot attendants, chairlift operators, concession and food service workers. It was even easy to find alpine rangers, who acted as ski patrol on the groomed slopes of Badger Pass. But Nordic, or cross-country, skiing—despite a centuries-old history—hadn't taken off in the United

States to the degree of alpine, or downhill, skiing, so few applicants possessed the winter survival and mountaineering skills required for the job.

Smith had to fill two positions, and about a dozen applicants made the "maybe" stack. After interviewing and skiing with them, Smith settled on Joe Evans, a seasonal ranger in his midtwenties who seemed to fit the mold of the new breed of ranger Smith was looking for. Evans was gung ho. He had aspirations for a career with the NPS, was an excellent skier, loved recreating in the outdoors, and displayed the right amount of adrenaline to do the job. "You could sense Joe was excited about the position," remembers Smith. "He ended up being the type of guy who would be waiting to join a search-and-rescue operation before I'd even heard there was a missing person."

The other ranger wasn't such an easy decision. Granted, Randy, at 32, could parallel-ski downhill on his three-pin cross-country skis better than most of his alpine patrolmen, he was already considered a veteran backcountry summer ranger at Sequoia and Kings Canyon, he had climbed in the Himalayas, he had been skiing the area he'd be patrolling since he was a teenager, and, offsetting Evans, "Randy seemed remarkably calm," says Smith. "But he wasn't a 'company man.' He was living what we called back then an 'alternative lifestyle.' Not into drugs or anything like that—he was completely content chasing the seasons, being a seasonal backcountry ranger winter and summer and taking photos for his aspiring photography career. He also asked if I minded if he kept his camera with him and took photos on his patrols."

Despite his reservations, Smith could not, in good conscience, hire a ranger who was more career-oriented but had lesser mountain skills than Randy.

The Nordic uniform of the time was a black wool turtleneck under a gray shirt and woolen green knickers fashioned from a pair of long pants. So that was the image of Randy and Evans striding across the frozen Yosemite tundra toward Glacier Point or Dewey Point. When they came across a skier with a "flat tire"—a broken ski binding or ski tip—they'd fix it. If someone had twisted a knee in a tree well, they'd make the skier comfortable with a wool blanket they carried in their packs and pull him or her out of the mountains on a sled. And if there was a situation in the farther reaches of the park, they were the first called to be helicoptered in, because, as Smith explains, "they were the two rangers in the park who could survive out there if the helicopter couldn't make it back out because of weather."

On one occasion, Randy and Evans were helicoptered into the back-country near Triple Divide Peak. A plane had crashed, it was windy, snow squalls were settling in over the higher peaks, and just as they hopped down into the swirling snow, the helicopter took off with their packs still inside. The two rangers looked at each other, then at the gray mountain tempest that was moving in, and said, "Oh shit." They had no survival gear. Fortunately, the helicopter was able to make it back through the clouds to pick them up after the operation.

During his tenure as a Yosemite Nordic ranger, Randy was called to his first recovery, another plane crash. He was one of the rangers charged with extricating the bodies from the wreckage—a grisly task in any season, but particularly haunting against the cold and snowy landscape of winter.

Randy took that first recovery in stride, though he admitted to Judi that it was the most difficult thing he'd done as a ranger. "The first one or two recoveries are the toughest," says Evans. "But the job had to be done, and if it wasn't a child or other tragic incident, gallows humor usually applied." Phrases like "Well, shit happens" and "By Gawd, every American has the right to die in their national park!" were lobbed back and forth. There was no post-incident counseling. "Back then, we did the job and went and had a few beers," Evans says. "In fact, it was not uncommon to have a few beers at the SAR cache when repackaging ropes and gear after the incident. Of course, Randy, being a bit older and in a serious relationship with Judi, was not always part of the 'faster crowd' of rangers at the time. Randy was more philosophical about life than most."

But if you got a little wine or beer in him, you couldn't get him to stop talking. Randy would crack Evans up with his sardonic sense of humor in one sentence and ground him in the next. "He reminded me to appreciate life and the wonders of the natural world," says Evans—and he wasn't afraid to explore the mountains in the winter, which "wasn't being done much back then." That winter, Randy recruited Evans and their friend Howard Weamer to ski from Yosemite to Mammoth. Nobody, to their knowledge, had taken the route they pieced together on a map, which began at Ostrander Lake and continued along the Merced Pass ridge to Banner Peak (which they climbed). On the fifth day of the trip, they made it to Mammoth. "With Randy and Howard, I am sure I came close to dying a couple of times by skiing across too steep or unstable snow," says Evans. "The classic 'whump' of snow settling still rings in my ears as I reflect on that trip. A splendid time, though, and we were blessed with perfect weather."

ON JUNE 14, 1975, Randy was helicoptered to Crabtree Meadow with a season's worth of supplies, another blank journal, and a copy of Edward Abbey's *The Monkey Wrench Gang*. He also carried with him a mild degree of hope. During the winter he'd written a few more stories, but he still felt strongest about the one inspired by his trip to the Four Corners Monument with Judi almost two years earlier. Sensing that it was a timeless piece, he'd rolled a sheet of paper into his father's typewriter and pounded out a new version of the story titled "Four Corners—A Prelude." He sent it off to *National Parks & Conservation Magazine* after having previous versions rejected by six other magazines. He also sent it to Wallace Stegner, with whom he'd been embarrassed to correspond because he still had not gotten any of his stories published after nearly three years of trying. He could have wallpapered his room with rejection slips and letters.

Before he went into the mountains for the summer, Randy asked his parents to watch for any response from Stegner or the magazine. He instructed them to have their mail forwarded to Sedona, Arizona, a growing artist community where Dana and Esther were overseeing the construction of the home they intended to retire in. Randy made it clear that if they heard anything, regardless whether it was bad or good, they should foward the mail to him in the backcountry or contact the park dispatcher, who could relay a message.

By season's end, he had hiked nearly 800 miles, spread the gospel to more than 1,400 people, evacuated a dozen individuals for medical problems—and heard not a single word from his parents.

Even without that letdown, leaving the mountains had always carried with it depression. By all accounts, Randy wasn't programmed for civilization. Something in his being short-circuited the minute he left the Sierra, disallowing total happiness. The first few days were always the worst, but Judi was there to ease the transition. Having recently graduated from California State University at San Jose with a degree in art history, she met him at Ash Mountain and drove with him north through the park, in the shadows of the Giant Forest. On the curving mountain road, Judi filled him in on her job hunt in San Francisco. Randy caught her up to date on the newly implemented wilderness-permit trail-quota system, which he wasn't entirely in favor of. He liked the idea of quotas—limiting the number of people allowed on a certain trail could only be good for the wilderness—but forcing visitors to provide an itinerary on a permit contradicted his idea of a wilderness experience, which, in his words, should be "impulsive and subject to change at the whim of the traveler." He understood the reason: to tally visitors and to track down overdue hikers who might be lost or injured. Still, he didn't like it.

Judi, though thrilled to be reunited with her man, was concerned that he wasn't even flinching at her mention of getting a job in San Francisco, a city, a place he would never consider living in. What, then, of their relationship? They loved their time together, but they were also comfortable apart, and comfortable "going with the flow" from season to season, year after year. And that, she worried, might become a habit she would eventually regret.

They got into Yosemite after dark—ten days past the full moon but it was still bright enough to highlight the granite walls rimming the valley. The valley always took Judi's breath away. "What a place to call home," she said as they entered Randy's parents' house.

Dana was still awake after having presented his regularly scheduled photography slide show to a crowd of 110 at the Ahwahnee Hotel.
Life was good for the self-taught photographer and botanist, who had
come to be one of Yosemite's resident celebrities.

At the Curry Company, he had worked his way up from managing
the accounting office to managing the reservations department, where
he was then promoted to director of reservations and, eventually, director of guest activities. The job was at first administrative, but soon
Dana was using his knowledge of the seasons, wildlife, and natural
world to act as a sort of public relations representative. When Randy
left for the Peace Corps in 1967, Dana was 57 years old. He had been
aiming for 60 as a good time to retire to Sedona.

His superiors, however, didn't want to see him leave, and in 1968
they created for him a unique assignment, previously untried by any
national park concessionaire. Combining two of his passions, photography and nature walks, with his natural ability as an orator, Dana
led camera walks for park visitors. The camera walks became one of
the park's most popular attractions, and he was so enamored with
the job, he extended his retirement to 1974, and then another year
to 1975. On this night, he let Judi and Randy know that he had decided to extend his retirement again, to 1979. He felt he owed it to
the public.

Letters to the Park Service praised Dana's skills. "Among [the park's]
fine human resources, there is one very special man, Mr. Dana C. Morgenson, naturalist, guide, lecturer, photographer, and author. [Dana
wrote and provided the photography for two books in the 1970s: *Yosemite Wildflower Trails* and *The Four Seasons of Yosemite*.] To take
a morning walk with the sensitive, artistic, knowledgeable, kindly
human being, enhances our appreciation of the Park's loveliness."

Another park visitor wrote: "Took the first walk [with Dana] on a
Wednesday and didn't miss one the rest of the week. His patience was
inexhaustible, his knowledge of the Park and its history, incredible.
The quiet trails along the Merced, with its shadowed pools held us
spellbound. Thanks to Mr. Morgenson, the Park really means some-

thing to us. The manner in which he described an incident in history for a particular view brings it to life. I noted with interest that all ages seem to derive the same delight that captivated us. A trip to Yosemite without a hike with Dana is but half a trip!"

His father's success motivated Randy: If you worked hard, believed in what you did, and stayed the course, then success and recognition would follow.

Randy and Judi retired upstairs to his old bedroom and, as was customary after the season, he pulled from the dresser drawer a stack of mail that his parents had collected. Toward the bottom was a letter from *National Parks & Conservation Magazine,* postmarked June 6, and another from Wallace Stegner, postmarked September 23.

The magazine had accepted his story—for the August issue. Now, in October, the story was more than two months late. Stegner had told him that once he got beyond publishing that first article, credibility would help carry him through to the next, building momentum for his writing career.

Stegner's letter was bittersweet.

Dear Randy,

I enjoyed your piece on Four Corners, and am surprised that you haven't placed it somewhere—though perhaps the contrast between then and now is a sort of inevitable subject, and has been done several times. I myself did a similar sort of piece on Glen Canyon as river and Glen Canyon as lake, and I remember most vividly the kind of contrast you speak of in your piece.

But whatever luck your piece has had, I'm moved by your feeling for the untouched country, the sand and the ledges and the sparse millet and the air, and the distances. Keep at it, it'll jell one of these days. And sometime go up on the Aquarius above Torrey, and climb to the very rim, at above 11,000 feet, and look off eastward. If the smog . . . hasn't gummed it all up, and it probably has, you can see the San Juan Mountains clear over in

Colorado, and a couple of hundred miles. I don't know anywhere
on Earth where you can see that far, except that view from that
high rim across that eroded desert to another high rim. But I
guess I wouldn't go there now unless after a cleansing rain or a
windstorm . . .

Yours,
Wallace Stegner

Judi knew Randy was crushed. "He'd worked so hard on that story," she says. "It was as perfect as he could have gotten it." He lay in bed, fuming at his parents. It tortured him to think that all that was needed was his signature to okay the $50 fee the magazine had offered to pay for his story.

Being the good son who rarely raised his voice in anger to his father and never to his mother, he didn't let on to his parents the full extent of the blow. He followed up with the magazine, but it was too late. They'd covered the Southwest fully in that issue and wouldn't be publishing stories on that region for a number of years.

JUDI RETURNED TO SAN FRANCISCO a week later to continue her job hunt. She hadn't voiced her concerns over the relationship with Randy, but she had shared them with her roommate, Gail, who, since the seventh grade, had been her closest friend.

Randy called Judi shortly after she arrived. The conversation barely got started before Judi was talking about the drive back and forth from the city to Yosemite; how she could and would continue to do it, but that she needed something more concrete than the good times she and Randy shared. She could barely believe her own ears when she heard herself use the word "married" in a shit-or-get-off-the-pot tone.

She was met with momentary silence, then Randy calmly said, "Okay—then let's get married." Judi and Randy had proven themselves a good match: neither of them wanted children; they shared similar philosophical and political views; they were content apart and happy together. Holding her hand over the phone, she told Gail quietly, "He just proposed. What should I do?"

Gail—who liked Randy and, more important, liked Randy with her friend—nodded her approval.

On November 22, 1975, Judi and Randy were married by a justice of the peace in a short, nonreligious wedding. The backdrop to their vows were Half Dome and the golden-grassed Ahwahnee Meadow, where Randy had played as a youth.

"This was the wedding day for Judi and Randy," wrote Dana in his diary. "The entire company walked out toward the sunny meadow in front of our house for the ceremony, which took about ten minutes. It was quite nicely done and a pleasure to participate in and see. Judi looked very pretty in her wedding gown, while Randy too was handsome in his blue leisure suit. Then everyone returned to the superb buffet Roy and Dottie Douglas had prepared and with plenty of champagne the occasion was a properly gay one."

Dana and Esther treated the couple to a honeymoon in the Gold Country, north of Yosemite. Randy and Judi drove up Highway 49, exploring the towns born from the California gold rush and staying in the romantic bed-and-breakfasts that caught their eye along the way.

CHAPTER SIX

PAID IN SUNSETS

Carried away a pack full of the leavings of Swinus
Americanus. Slobs, creating their own bad karma, give
me the chance to do Earth a good turn. Perhaps blessed
stormy weather, and succession of rangers, and god will
have the joint looking natural in 10,000 years or so.

—*Randy Morgenson, location and date unknown*

I find that in contemplating the natural world my
pleasure is greater if there are not too many others
contemplating it with me, at the same time.

—*Edward Abbey,* Desert Solitaire

ALDEN NASH TRANSFERRED to Sequoia and Kings Canyon from Yellowstone during the winter of 1975. As the new Sierra district ranger in charge of most of the backcountry rangers, Nash spearheaded a movement that helped nudge the parks' policy out of the dark ages.

For decades, the recruitment of backcountry rangers had been what some considered both chauvinistic and militaristic. Nash, the father of two daughters, couldn't see "any reason whatsoever" why a young woman could not do the job. Before Nash, a backcountry ranger had a better chance of being hired if he was a "clean-shaven white boy without a girlfriend," a statement Nash follows with "I'll deny that in court."

Nash changed policy at Sequoia and Kings Canyon abruptly when he hired the parks' first female backcountry ranger, Cynthia Leisz, who had been working in the frontcountry and was capable and keen to break the mold.

Simultaneously, Nash toned down the dress policy as well—an impulsive, unauthorized move instigated during his first meeting with Randy. In early June of 1975, Nash walked into the district office at Ash Mountain and found a fully bearded man bantering with the secretary. The secretary looked over at the official-looking Nash, who was clean-shaven and wearing a pressed uniform, badge, and the traditional ranger flat hat.

"Ask him, he's your new boss," she said.

Randy turned around and faced Nash, who, at 33, was the same age, and introduced himself: "Hi there. I'm one of your backcountry rangers. Any chance I can keep this beard for the season? It helps keep the mosquitoes off."

The regulations dictated military-style haircuts (above the ears and collar) and absolutely no facial hair beyond a neatly trimmed moustache. Still, common sense told Nash the regulations were outdated. Why shouldn't a mountain-man beard be allowed in the mountains? But maintaining an air of authority, he countered Randy's bullshit "mosquito" line with a straightforward "Is it going to affect the way you do your job?"

Taken aback, Randy responded, "Not at all."

"Then I don't see why you can't keep it," said Nash.

A huge grin burst through all that hair, and Randy introduced himself properly with an enthusiastic handshake, quickly adding that he had been the Crabtree Meadow ranger the season before, and if at all possible, he would like the same duty station again. "There is some housekeeping I didn't quite finish last season that I'd like to follow up with," he said.

Nash, thinking that perhaps the cabin's roof needed repairing, asked, "Something wrong with the Crabtree station I should know about?"

"No, no," returned Randy, "housekeeping as in cleaning up after the campers—illegal fire pits, busted-up drift fences, things like that. The cabin's fine."

Nash knew he was going to like this guy.

For Nash, that first summer as the Sierra district ranger was a dream come true—a homecoming of sorts. He started his career with the Park Service as an 18-year-old seasonal firefighter for Sequoia National Forest in 1961. In 1962, he worked the backcountry on a soil and moisture crew, building check dams and cutting down small trees in overgrazed meadows. In 1963, he worked on the helicopter firefighting crew at Ash Mountain, and in 1964, the year he graduated from Humboldt State University with a degree in forestry, he was a fire control aid at Ash Mountain. He then spent eight weeks in basic and advanced infantry training for the National Guard before landing his first permanent job as a ranger in Yellowstone.

To say that he knew the ropes of the National Park Service was an understatement, but he didn't come close to knowing the backcountry of Sequoia and Kings Canyon as well as the rangers he was supervising.

Before the end of June, a freak snowstorm left 4 to 8 inches of snow on the ground across the high country, making it a challenge to get over the high passes that separated Nash from his rangers. For Randy, it was a welcome act of nature that he thought might deter the first wave of backpackers, giving him a few more days to soak in the solitude. That wishful thinking didn't take into consideration those who were already in the backcountry and trapped by the storm.

Sure enough, a half-dozen "guests"—unprepared, shivering backpackers—kept Randy company like sardines in his little cabin the night of the storm. The next day, he kicked his "guests" out as quickly as possible so he could lock the place up and go on patrol in the unseasonable winter wonderland.

Not long after the storm had blown east, off the crest, Nash radioed Randy to let him know he'd be stopping by shortly. "You realize the passes aren't open to stock yet, don't you?" said Randy.

"I won't be on a horse," replied Nash.

This impressed Randy, who, in his seven seasons, had formed a semi-contemptuous opinion of horses and mules in the backcountry, along with some of the men and women who rode them. He'd also formed an opinion of administrators and supervisors—most of whom, he observed, rode stock or helicopters.

Randy had seen only one administrator in the backcountry on foot: in 1974, when he had met Bob Smith, the chief ranger at the time, backpacking in Randy's patrol area. "This is the first time I've seen administrators hoofing around out here," wrote Randy in his logbook, "so I congratulated them. Don't think they knew just how to take it."

A different breed of district ranger, Nash refused to let go of the wilderness fieldwork that so often was lost under piles of paperwork. Generally, the shackles to a desk clamped down the second a ranger was promoted to district ranger or chief ranger. "Heaven forbid a superintendent wear out a pair of hiking boots," said Randy, who was perplexed by the bureaucracy. The higher up you got, the less time you were expected—or were able—to spend in the wilderness you were supposed to be managing. "What's the incentive?"

Nash, who carried a mini office in his backpack, took advantage of any spare moments while in the backcountry to stay on top of the mountain of memos, employee evaluations, incident reports, and other paperwork required of a government employee in a management position. He also learned the favorite fruits and vegetables of his rangers, carrying in salad fixings, fresh green beans, once a whole watermelon. And he gave many long pep talks while hiking with Randy and other rangers. They weren't making much money, and he felt that a few pats on the back might keep them coming back, which was good both for him and for the "resource"—a government term for its parks. Randy preferred to call it "the country," reasoning, "Would you use the word 'resource' to describe your wife?"

Nash conveyed to Randy that backcountry rangers were like scouts, or forward observers, who reported back to higher-ups what was going on in their territory. In that capacity the rangers were crucial in their roles in managing the wilderness.

Randy was receptive, but let Nash know that he didn't think his and the other rangers' voices were being heard.

Nash made the same promise to Randy that he made to all his rangers that season: "I'll listen."

IN THE SPRING OF 1977, during a month he had off between his winter and summer ranger duties, Randy took a long-dreamed-of trip to Alaska with Chris Cox, a buddy of his who was a climbing and ski guide in Yosemite. On this mini kayaking expedition to Glacier Bay, they camped alongside the thunder of an eroding glacier, paddled around icebergs, explored the islands and glaciers John Muir had written about, and barely slept while using a campsite where, the year before, a man had been eaten by one of the resident coastal brown bears. They watched bald eagles plucking fish from the sea, newborn seals basking on rocks, whales in their wakes, mosquitoes big enough to make off with a small child, and put their lives in the hands of a charter pilot who guessed that his oil-dripping plane could "probably handle the weight" of their gear.

Randy arrived home in Yosemite with less than a week in which to buy his food and supplies, get down to Sequoia and Kings Canyon, and hop a helicopter to Little Five Lakes, where he was stationed for the summer. There was little time for Judi, who in Randy's absence had been hired by famed mountaineer Ned Gillette to work at the Yosemite Mountaineering School—a full-time summer gig catering to the throngs of outdoor enthusiasts who came to the valley to experience the school's motto (and best-selling T-shirt), "Go Climb a Rock." Randy and Judi had been married for a year and a half, and more than a third of that time they'd been apart. A friend had told Judi that they were living a military lifestyle. After some contemplation Judi agreed, but was quick to explain how she didn't really feel the separation because their connection was so strong. They made it a point to communicate as often as they could, and she had—each summer since they'd met—hiked in to see him for sometimes weeks at a time. This year, however, would be different.

She broke the news to Randy that she wouldn't have enough time off from her new job that summer to hike in and see him; it would take two days just to get into his patrol area. Randy's response was hurried and along the lines of "So, I won't see you till the fall? That's how it's going to be from now on?"

Judi was taken aback. She was irritated by Randy's attitude: he seemed to think she should drop everything to go and visit him in the backcountry, even though she had a job that she was looking forward to. "Well, if you wanted to see me," she said, "maybe you should have left some time after Alaska, before you had to go into the mountains."

She went on to tell him that she was going to be taught basic rock climbing at some low-angle granite not far from the mountaineering school so that she could speak the language while working the school's counter. Judi was excited to be learning rock climbing, not only because of her own increasingly adventurous spirit but also so that she could feel more compatible with Randy, who was—since guide school in the Himalayas—comfortable on vertical ice and rock.

Climbing was scary, but once the instructor showed Judi the rope techniques and demonstrated how he would stop any falls, she began to enjoy it. A couple of routes later, every muscle in her hands, arms, forearms, and fingers was taxed, but she was having fun. After down-climbing a pitch on a rock she never in a million years thought she could have handled, she hopped down and gave her instructor a hug born from the excitement of the accomplishment.

"It was straight out of a movie," says Judi. "I jumped off the rock and gave my friend a hug the second Randy came around the corner and saw me in this guy's arms."

It was innocent but awkward, and Judi forgot about it immediately. After all, Randy was known to flirt and, being a local to the valley, often hugged other women in greeting. Judi was confident in their relationship.

Randy, however, was not.

At Little Five Lakes, Randy wrote in his personal diary, "The last few days together before my summer here, jammed with the presence of friends and my need to collect equipment and shopping lists for the backcountry scattered my energies like pollen in the wind, giving too little to her and too much toward others. . . . But a little distance, and a look back, and what was important was Judi and her place in my life, how much I love her and how I need to cultivate that, direct my energies there."

In the solitude of the backcountry, he obsessed—rangers call it "looping"—noting that Judi's usual playfulness hadn't been there, as well as "a distance," he wrote. "A distraction. An expression in her eyes and on her face which said, 'Randy, I'm sorry, there are things I'm not telling you.' And it was left at that. Hanging."

Weeks later, he was still analyzing, this time about the very moment when they'd said goodbye: "Don't worry about me, she said, the last thing, with a tone and a look of very un-reassuring reassurance."

A dozen pages later in his diary: "The thoughts don't stop. The things that seemed to indicate a difference, did they really or am I working them up? Several times I mentioned there seemed a difference about her, finally saying it as an invitation to talk, to say something, to give me a clue so I wouldn't be left wondering blindly all summer. But her only response, 'Oh? How?' No denial, no explanation, no comment. And around the mountaineering school she seemed more interested in the people there than in me—at least more lively with them. . . . Am I conjuring it all up?

"My poor feverish imagination. This summer, who is she sleeping with? How often? Where is her love? Is there anything left of us? If our young marriage survives this summer I'll need to give her more love, keep her closer, make us a real couple, keep a love going with us, forget my flirting ways with other women, and end these long separations— that above all. Judi shall be in the backcountry with me next summer. This ruins a mountain high.

"These feelings about us I wanted her to understand before this backcountry summer began, but was cut off. 'This is all crazy. She'll be there. It'll be good. We'll have a good winter in Tuolumne,' I tell myself."

Five days later: "No letter. Shit. More than a month has passed."

That evening a stock party camped in a nearby meadow, which mercifully took Randy's mind off his worries. For two full pages Randy was back to his old self, defending meadows and their processes from the evils of man. "The meadow isn't here to make a comfortable camp-site for you," he wrote, "so don't circle rocks upon it and build your fire there. Nor is it here to provide feed for your horse's belly. Be respectful. You are on holy ground. Step lightly. Keep your imprint, your intrusion, your 'use' to the barest minimum.

"This has been written about for decades by the New England Tran-scendentalists, by eloquent writers like Wallace Stegner. By cantanker-ous writers like Ed Abbey. So here I am trying again. How many are being reached? How much progress against the machine mentality are we making?"

Finding companionship in his surroundings, Randy wrote, "These things I watch year after year, and take leisure in knowing the names of these sedges and grasses as I watch them go through their changes like knowing the name of a friend whom I've known through the years.

"In Alaska, I became excited about the newness of the country. My curiosity became piqued and I yearned to know the plants, geology, glaciology, weather, and the effects of these things on each other. I felt like a stranger. And returning to the High Sierra, I realized the com-forting feelings in knowing many of these things. Like being among friends."

Still, Randy was lonely, and he missed Judi. Ever the thinker, he justified his emotions by reasoning, "An element of loneliness is neces-sary in wildness."

IT WAS THE DAY after Thanksgiving 1977, and Randy and Judi had survived what Judi describes as their first "rough bump" on the road of marriage. Randy never let on how worried he had been at the begin-ning of the summer. Judi was, as she put it many years later, "blissfully content" and, as the snowplow rumbled past on the road below their cabin in Yosemite, "excited."

They were snowbound—together—for the rest of the winter. Randy had been hired as the full-time Tuolumne Meadow winter ranger. Judi, who, thanks to Randy, had learned Nordic skiing herself, was hired part-time, two days a week. They were a team, and now with the Tioga Pass road officially closed, they were "alone with the storm and the mountains," wrote Randy. "The way we want it."

Three feet of snow had fallen when Randy and Judi woke to a "Clear. Sunny. Brilliant" day. "Impossible not to go outside for a ski," wrote Randy that night inside their cabin—a barely insulated two-room affair that had been drafty before they patched the walls with cardboard cut from the boxes of provisions they'd brought in for the winter. "Snow has drifted around the back door, collapsing into the room as I first open it, but after a little shoveling we put on skis in the doorway and pole off the porch.

"Breaking trail. Forcing a way, with the snow pushing back with an equal and opposite force. Elementary physics, but who cares? The world is ours and as beautiful as it can ever be."

When the road was still open, Judi had brought in—with full National Park Service blessings—her "Japanese clothes dryer": a kiln for firing the ceramic pottery she intended to craft that winter, between the shoveling of snow and the ski-patrolling up and down the Tioga Pass road and the other routes threading through the snowbound and deserted park.

More than 30 feet of snow would fall that winter—a constant battle of clearing rooftops and pathways to the outhouse and trying, without success, to maintain a ski track, at least around the meadow and a fair distance in either direction on the road whose blacktop wouldn't see wheeled traffic for more than five months.

For a young couple in love, it was a dream job.

"We push through to the lower end of the meadow, to Pothole Dome, leaving behind a set of tracks which should be a pleasure to ski on our return," wrote Randy after the second major storm. "There's no wind, and the sun ricocheting off the ambient white is hot. Judi takes off her sweater and shirt, to ski au natural. Sweet nakedness. I suppose

I should take off my pants and likewise flap in the morning sunshine, but I elect an ascent of Pothole Dome instead, leaving her bronzing her body on our ski track."

Fresno Bee reporter Gene Rose learned of the husband-and-wife team, together in the high lonely of the Sierra for an entire winter—as romantic a hook as he could imagine. He interviewed them on the telephone, their only link to civilization, about their snowbound winter jobs. Rose's editor buried the article in the back of the paper.

In spite of that, United Press International's wire service picked it up and followed with its own story on the couple the next week. NBC from Texas and NBC from Burbank, California, called, both wanting to fly in via helicopter to interview the rangers for the news. Rose's editor apologized for the lack of foresight; apparently backcountry rangers were worthy of a story in the front of the paper.

Meanwhile, Yosemite's park information officer called the Morgensons to tell them she had been fielding calls from New York, Alabama, Florida, and on and on. Judi was interviewed by a man from KNX, Los Angeles. On February 23, CBS "called us to tape an interview for a 'woman's related program,' " wrote Randy. "Their most earnest question was about how we get along living so close together for so long. Don't we fight, scream, and tear each other's hair? Of all the things she could have asked about our life here . . ."

Randy mused upon the things he would have asked, and what he and Judi could have told: how taking a bath was an all-day affair— digging out the bathhouse, melting snow, firing the tub, keeping the fire stoked every twenty minutes, more snow, more wood. Meanwhile, clothes would be soaking in the cabin, the "soak cycle." How the meadow at dusk presented a constant show: "a few thin streamers of rose pink, and purple clouds in a hard, cold blue sky; a thin wash of the lightest orange on the mountains, and the evening mist, thick and cold, rising off the meadow." How they could spend a morning watching the heavy snowfall while huddled around a warm waffle griddle. About the miles and miles of country crossed to the squeaky tune of waxed skis on fresh powder, "exhilarated by the winter world we moved through. The pines and firs, some of them giants among their kind, were coated, every branch, twig and needle—the entire world was an even white." How sometimes they would not make it back to the cabin until well after dark, retracing their tracks in the moonlight or by flashlight "up our hill to our reward—hot buttered brandies." How the evening's entertainment might be listening to *The Nutcracker Suite* or reading Annie Dillard's *A Pilgrim at Tinker Creek*. About the uncountable cups of tea. The cast of woodland creatures. The mouse that washed its face and head "just like you, Randy," laughed Judi one night. Or the pine martens—a male and female—that became their favorite to watch. "Like flowing water, he flows over his terrain," wrote Randy. "There are no obstacles. Back slightly arched, alert eyes on the snow in front of him, tail moving with slow undulations behind, and furred feet running rapidly over the snow surface he moves with an almost effortless grace. Whenever I watch such a creature in the wild I think, with amusement, of how awkwardly we move about within our medium."

Most importantly, Randy would have told the audience why they were there in the first place: as a safety presence for the few wilderness travelers hardy enough to ski into the area, and to monitor the park and the wilderness overall, which meant maintaining the buildings and a near-constant battle of shovel versus snow.

Too boring, Randy reasoned, for television and radio.

One day during this media bombardment, they heard boots on the step and a knock. There in the doorway, like an arctic explorer with red hair and eyebrows frozen beneath a hood, was Randy's good friend and fellow ranger, George Durkee, come to visit Yosemite's "latest and greatest celebrities."

Durkee had begun his career with a U.S. Forest Service fire crew in 1970, but landed a position with the NPS in Yosemite in 1973 before being recruited to Sequoia and Kings Canyon in 1977. Shortly after his recruitment, he was hanging out in the valley in full hippie attire—including ripped-up denim jeans with matching jacket and bandana headband—long-haired and suspect outside the Curry Village grocery store. A high-level ranger from Sequoia and Kings Canyon named Paul Foder was walking by with one of the Yosemite Mafia. He pointed toward Durkee and said, "Now, that looks like probable cause if I've ever seen it."

The Mafia member said, "No, that's the new ranger you just hired."

Durkee would never live that one down. He'd originally been hired as part of the effort to better relate to the youth population after the riots—which meant getting out and walking the paths and camp-grounds of Yosemite at night, "getting the youth to fly right," says Durkee. One night as he was walking in Stoneman Meadow—without a flashlight because he was "extra cool"—he came upon a dark form just off the path.

"Hey, guy, sorry but you can't camp here," he said.

No response, so a bit louder, with authority: "Park ranger. You can't sleep here. You've got to move."

Nothing. Durkee cautiously kicked the guy lightly with his foot. "Klunk."

It was one of the cement forms used to separate the trail from the meadow.

He slunk quietly away, only to be reminded of "the incident" by Randy—repeatedly—after Durkee made the mistake of telling him the story. This night, as he stood in the doorway of the Morgensons' cabin in Tuolumne, was no exception.

"Hey, Judi, it's that ranger who tried to arrest a cement pylon," said Randy before giving Durkee a hearty handshake and pulling him out of the cold.

The two friends were sardonic, cynical, hypercritical, anti-establishment, irreverent lovers of wilderness who bantered back and forth inces-santly. Randy would yawn when Durkee quoted his favorite authors—Herman Melville and Joseph Conrad—while Durkee would "tolerate amateurs" like Randy's favorite, Thoreau.

Durkee, Randy, and Judi ate dinner and spent the evening discuss-ing the now-required Law Enforcement Commission for backcountry rangers, theorizing that during the classic police-academy "defensive driving" segment they would set up deer and backpackers to maneuver around. Since they didn't drive squad cars in the backcountry, they'd just run "really, really fast" through the course while making driving sounds—screeching tires and the like.

"Company like George was always a treat," remembers Judi. "I'd just sit back and listen to them like they were Abbott and Costello. They were the entertainment."

Durkee and Randy's friendship would be strengthened over the years by a dirty little secret—an addictive, mutually perpetuating hobby of bashing the National Park Service. They also had a propensity for looking out for their fellow backcountry rangers.

One of their first victories against the NPS was getting rid of the rent rangers were charged for their backcountry stations for most of the 1970s. These rudimentary shelters, loosely termed "cabins," generally had sparse furnishings, no running water, no electricity or plumbing, and were infested with mice, rats, and the occasional porcupine. Durkee and Randy felt that their canvas tents and drafty, leaky cabins weren't in the same league as the "public housing" in the frontcountry, which was guarded by "private residence" signs.

There was nothing private about a backcountry ranger's station. In addition, they were used by the public as emergency shelters during storms and as trailside motels for administrators passing through. And there had to be some underlying reason for the free utilities. Perhaps it was a way around shelling out hazardous-duty pay for having to light combustible cooking stoves. The amount of rent charged was reasonable enough—$14 a month for the Little Five Lakes tent, $21 for the Rock Creek cabin, and $15.17 for the stone Tyndall Creek cabin. But Durkee and Randy felt it was their duty as Americans to point out the injustice. It was the 17 cents that pushed them over the edge.

Visits from friends like Durkee to the Morgensons' Yosemite duty station were the exception, not the norm. Their company was each other and the woodland creatures that weathered the winter. And then the winter that had been before them was melting in their wake. The rivers grew louder, the bears staggered groggily out of hibernation, and the wilderness yawned and slowly awakened. Throughout it all, the Morgensons had confirmed their compatibility: how their interests could sustain them and how life, in its simplest form, was entertainment enough. Every day they'd discovered something new, another example of nature at work. Skiing across the meadow after a fresh snow, they often came across a rodent's tracks. If they followed these tracks, they would sometimes end at a burrow in the snow, with icicles formed over the opening from a mammal sleeping inside. But one day the tracks suddenly became erratic and disappeared.

Stumped, with no clue as to what had become of the rodent, they stopped. Then they noticed a perfect "snow angel" at the terminus of the tracks—indentations made by an owl's wings as it pounced on the animal from above, pushing its small body into the snow. A few specks of blood were the only sign of the kill. The discovery made Judi sad, but Randy explained it as nature in its rawest form—a celebration, if not for the mouse, for the owl. This was a world Randy understood and the kind of life in which he felt most comfortable.

"Back in civilization I begin the questioning," wrote Randy. "What to do with life? What kind of life? In wilderness this ceases; the questions aren't answered, they dissolve."

BEFORE LEAVING for the required ranger law enforcement academy in Santa Rosa, Randy went to Yosemite's wilderness office to see his supervisor, a subdistrict ranger who had checked in on the Morgensons a few times that winter and brought some VIPs on one visit. Randy's intent was to review his performance rating, then sign to confirm that he agreed with his supervisor's comments. He was informed, however, that his supervisor hadn't had time to complete the review, or even to start it, for that matter. Confident in his performance, Randy signed a blank document.

While Randy was learning how to be a "wilderness cop," things weren't going so well back in Yosemite. Dana Morgenson had his annual physical, during which a suspicious lump on his prostate was diagnosed as cancer. On May 29, as Judi shuttled gear to Dana and Esther's house in the valley, she developed a headache and nausea, which she at first attributed to "reverse altitude sickness." On her second trip, the headache became so painful that "it was hard to see straight."

Judi ended up hospitalized for four days with aseptic meningitis, described to her as a less deadly form of viral meningitis. Doctors theorized that she had contracted it from the deer mice they'd lived with all winter, one of which had bitten her. The park refused to pay her hospital bills, stating that "illnesses" contracted while on duty weren't covered. She was told that if she'd broken a leg, it would have been a different story altogether.

Despite the distractions, Randy earned his law enforcement commission.

Back home in Yosemite, the family celebrated with a steak dinner. Randy checked his dresser drawer for mail and found a letter from Yosemite National Park: his winter's performance review.

To his dismay, there were low marks across the board and, toward the end of the letter, a seasonal ranger's worst nightmare: "Do not recommend for rehire." The main reason was something to the effect of "The ranger's lack of motivation to patrol on skis the area assigned, failure to shovel snow-loaded roofs." Worst of all, there was Randy's signature confirming that he had read and agreed with the report.

Randy and Judi surmised that they must have offended the supervisor when he'd showed up with the group of VIPs. It was a short-notice visit that came not long after a huge snowfall. The ranger had skied in without a broken trail, which was, perhaps, the first strike. Of course, before the storm had begun, there was a perfect track: not a week went by all winter when Randy didn't circumnavigate the meadow once or twice. Then the supervisor had suggested that Randy come out for a ski "with the boys" and Judi stay behind to prepare them a meal. Randy had refused, offering in return Judi's guiding skills or his cooking skills. They skied alone, and Randy greeted them at the end of the day wearing an apron.

Randy and Judi compiled a list of duties from the winter: they'd encountered 202 people, all on skis; they'd documented hundreds of miles skied; for each day he hadn't skied, they'd performed some "housekeeping" duty, including relentless snow removal from rooftops. In addition, the days he hadn't skied, it had snowed—heavily. There appeared to be no grounds for a poor review that squelched any chance of rehire for the position in the future and, worse, permanently scarred his perfect performance record with the National Park Service. He was going to fight the NPS on this one.

But attorneys were expensive—potentially more money than Randy and Judi made together in a year or more—and cases against the government weren't popular with lawyers.

Then Randy told the story to Durkee, who said, "Wait a minute, check this out." He pulled from his wallet the business card of San Francisco attorney Richard Duane, whom Durkee had assisted in the backcountry the season before when Duane's back had given out. As Duane lay on the shores of a lake in pain, Durkee had sat with him and chatted for hours, then took much of the weight from Duane's pack so he could make it out of the mountains. In parting, Duane gave Durkee a card. "If you ever need an attorney, give me a call," he said.

Duane took Randy's case pro bono.

More than a year after Duane confronted the NPS about the questionable performance review, it was struck permanently from Randy's record in an out-of-court settlement of sorts. Randy wasn't looking for money; he just wanted the bogus review removed.

Despite the victory, Randy was so embittered by the experience that he chose never to seek reemployment as a winter (or summer) ranger in Yosemite. Sequoia and Kings Canyon were his home parks, for good.

RANDY HAD ALWAYS told Judi that he would be happy living his life in a tent, but she knew he wasn't entirely serious. He'd learned to appreciate certain creature comforts: clean sheets, a cooked meal, a companion, and a lover.

During the spring of 1980, they bought a 700-square-foot house in Susanville, California, in the foothills of the eastern Sierra. Susanville was home to Lassen Community College, which had one of the best photo darkrooms in the state and a reputable art department for Judi, who was interested in instructing.

Randy berated himself repeatedly as they moved in, saying things like "So, does this make me a suburbanite?" But he soon got into the swing of things, setting up a photo darkroom, organizing the extra bedroom into an office, and cultivating his own private meadows, which other neighbors mistakenly called "yards."

Randy had an ethical problem with cutting grass. Judi tried to explain the difference between domestic grass and wild grass, but Randy resisted tooth and nail. One day when the grass was blowing in the wind, a neighbor leaned over the fence, pointed at the knee-high "lawn," and said, "You know, you're supposed to mow it." That was the front yard.

The neighbor off the backyard had a problem with dandelions, which blew in like sorties from Randy's meadow. He would comment loudly, "I wonder where all these *weeds* are coming from?!" For Randy, there was no such thing as a weed. He'd hear the neighbor and grin at Judi as if his master plan was working.

When it came time for Randy to return to the backcountry, Judi reminded him of the promise he'd made to her. With a wink, he drove off and a couple of hours later returned and unloaded the "promise" onto the driveway. Judi eyed the lawnmower suspiciously, and even circled it a couple of times for effect, before asking, "Where's the engine?"

"But that was Randy," says Judi. "He refused to buy anything but a push mower." Randy headed off for the mountains, and Judi bought a pair of leather gloves and began the ritual of maintaining peace with the neighbors by keeping the yard well manicured—for five months of the year.

On June 18, 1980, after "ten nonstop, world-class marathon days of training," Randy was flown into his duty station at Tyndall Creek.

THREE WEEKS LATER, 27-YEAR-OLD Nariaki Kose entered the Cedar Grove ranger station seeking a wilderness permit. Kose was a Japanese citizen who had been in the United States studying as a graduate student at the University of California, Berkeley. He had come to the parks to climb a series of peaks on a difficult, predominantly cross-country route that crossed the Sierra from west to east, from the Cedar Grove area to Mount Whitney. His intent was to make the trek solo.

An experienced mountaineer, Kose was described as "almost militaristic" in his lightweight techniques; for instance, he carried only one pint of fuel for the fifteen-day trip, which allowed for a single cup of hot tea per day. Anticipating snow, he did bring all the standard equipment—crampons, ice ax—necessary to negotiate steep, potentially icy terrain. But he likely hadn't expected the snowpack to be 180 percent above normal. The ranger who issued Kose's permit had recently hiked into the backcountry, and he warned Kose that all camping above 9,000 feet would be done on snow that, on north-facing slopes, was still 10 to 14 feet deep. Kose seemed receptive to the ranger's suggestion to change his chosen route through the Sphinx Basin to Bubbs Creek, then East Lake. At Reflection Lake, the ranger recommended Kose take a good look at the mountains he would be traveling in and decide then and there if he shouldn't come back and try it in August.

Kose had told a friend he'd be out of the mountains by July 24; when he didn't show, the friend contacted park headquarters. Backcountry rangers near his proposed route, including Randy, were alerted. No backpackers in the area had seen anybody matching Kose's description, but on July 27 a ranger at Mount Whitney, where Kose was supposed to exit the mountains, spoke with a person believed to have been the missing mountaineer. The rangers assumed that Kose was fine.

On July 28, Kose's friend called again. Kose was now four days overdue. A helicopter and ground teams led by Subdistrict Ranger Alden Nash were dispatched to search the proposed route. Meanwhile, hundreds of wilderness permits issued during the month of July were pulled. One group, whose route might have intersected Kose's, confirmed having talked with him on July 9 and 10. Kose had told the group that he intended to climb North Guard Peak on July 11. This information was radioed to Nash, who had joined Randy at his station.

As he usually did with mysteries like this, Nash turned to Randy and said, "Well, Randy, what do you think? Is he on North Guard?"

Randy consulted what Nash described as his "topographic-photographic memory" of the Sierra: the 13,327-foot North Guard is a couple of hundred feet lower in elevation than Mount Brewer to the south—a less-traveled peak that Randy had climbed himself more than once. Like so many Sierra peaks, from a distance it appears an impossible climb without a rope. But up close, the pockmarked tan-and-gray southern face presents some elongated scars—avalanche chutes—that are outlined in guidebooks as the most plausible route to its impressive, pointed summit.

Randy sensed that these chutes were likely what Kose had taken and, if he'd had an accident, the best place to begin the search.

The next morning, rangers in the parks' helicopter spotted a backpack a fair distance below the steepening face of North Guard. About a half mile above the backpack, Randy rendezvoused with rangers Ed Cummins and Ralph Kumano, and the three worked their way upward. "About 1030 or 1100," wrote Randy in his logbook, "I found his body in a gully 400 to 500 feet below the summit."

Some nineteen days earlier, Kose apparently had climbed to the top of this avalanche chute, to the point where a cliff section had to be contoured briefly before he gained the summit just 200 feet higher. While negotiating a number of boulders near the top of the chute, he had slipped and fallen more than 300 feet down the steep gully.

It was "particularly unpleasant getting Kose into a litter and out of there," says Kumano, who twenty-five years later still remembers what the Japanese climber was wearing: a red-checkered flannel shirt, tan gaiters, red-laced leather hiking boots. "I don't remember Randy saying a word through the entire process."

This was one of the first times since the Wilderness Permit System was implemented at Sequoia and Kings Canyon that it had been used successfully as a system to help locate a missing person. The regional director of the National Park Service commended the rangers for their tedious wilderness sleuthing:

The chore that you undertook to sift through the hundreds of Wilderness Permits . . . the one backpacker you located by this method who had met and camped with Mr. Kose was akin to the proverbial needle in the haystack.

The fact that your people were able to find the man's backpack and later his body is an accomplishment that truly verges on the incredible. More important, however, is the fact that the continuing day-to-day anguish of Mr. Kose's family and friends was undoubtedly greatly reduced.

One week after Randy helped find the body of Nariaki Kose, Dana Morgenson, who had finally retired at age 71 after thirty-six years working in Yosemite, trudged up the Shepherd Pass Trail to meet Randy in the backcountry for the first time since 1965. He joined his son at Anvil Camp, having climbed 4,300 vertical feet in 6 miles. Apparently all those wildflower walks had served him well over the years. He seemed in perfect health after having beat prostate cancer—a little winded, but in fine shape.

For the next seven days, father and son "walked in beauty," wrote Randy in his personal diary.

Had either of them known what would occur a month later, perhaps more would have been said. But their mutual interest in the surroundings—in the Sierra—seemed to keep personal conversation at bay. It had been that way since the Peace Corps.

If Wallace Stegner had been a family psychologist, he might have described their conversations as "big vague thoughts about big vague ideas." What they talked about lacked a certain depth—unless, of course, it concerned the scientific subtexts of a particular wildflower.

On a patrol from Randy's Tyndall Creek station to Lake South America, Randy shared with his father a magical campsite on a granite bench alongside a small lake—a spot Randy would return to time and time again. That day, it was significant because it symbolized an unspoken thank-you to his father, who "gave me these mountains when I was too young to understand," wrote Randy in his diary.

Likewise, Dana voiced his awe of the place by writing in his journal how morning had ushered in a "glorious day of wandering in lush, green meadows, by dashing mountain streams, past little blue lakes, with a dramatic skyline of snow peaks always in the background." They circled back via Milestone Basin, which Dana said was "as pretty a spot as I've ever seen." If Randy had a nickel for every time he'd heard his father say that . . .

Randy would agonize later in life over the things that were unsaid. He would accuse his father of not being there for him emotionally. And he'd accuse himself for his shortcomings in the relationship. Despite their near-perfect façade, the wilderness family Morgenson had issues. What family didn't? At the time, however, those issues weren't urgent—especially with his father and mother embarking on their long-planned trip to Alaska, where Dana would see some of the world beyond Yosemite that his hero John Muir had written about so eloquently—stories of glaciers, and storms, his dog Stickeen. Alaska was the stuff of a long-awaited adventure for the older Morgenson.

Although Dana had never been happy with his son's decision not to finish college, by walking with him that week he must have realized that Randy had found his true calling, seasonal as it was. But as far as he and Esther were concerned, their younger son was selling himself short by living the life of a seasonal ranger. "If anything, Dana thought Randy was too smart to be a backcountry ranger," says Jim Morgenson, Dana's brother. "Dana didn't really think Randy lived up to his potential, and I'm pretty certain Randy sensed this. It had been broached a few times, but I think it was a sore topic and avoided."

As the week came to a close, Randy escorted his father back over Shepherd Pass—an easy stroll for him, but Dana showed his years, however slightly, by inquiring, "Is that the top?" about another false summit. Or "How many more switchbacks to the summit?"

"Thirty years after my boyhood, the roles are reversed," wrote Randy in his journal. "Aren't we almost there, Daddy? How much farther?"

Randy bid his father farewell at Mahogany Flat.

"Peace," Dana said with a smile before turning and disappearing around a bend in the trail. In a state of contented melancholy, Randy strolled back to Tyndall, reflecting on one of the best times he'd ever had with his father.

Dana and Esther Morgenson embarked for Alaska on Friday, August 15. Two days after their forty-seventh anniversary, they departed from Juneau, Alaska, to the high point of their monthlong journey: Glacier Bay. The turquoise green waters at the head of the Reid Glacier kept Dana glued to the ship's railing. "Amazing," he wrote while they were moored off the snout of the Lamplugh Glacier. Then it was back down to the Lower 48, across the Juneau Ice Cap to Seattle, where Randy's childhood friend Bill Taylor hosted them at his house before they looped back toward home via the Olympic Peninsula. On September 19, Dana and Esther were camped in a rain forest. After dinner, Esther wrote postcards and Dana went to bed early. It was unusual for him to be in bed before Esther, and even more unusual for him not to punctuate the day by writing in his diary.

Early the next morning, Dana Morgenson suffered a massive stroke. He and Esther were rushed to Harbor View Medical Center in Seattle via Army helicopter, but doctors told Esther that life support was the only thing keeping him breathing. Dana had died on the Olympic Peninsula.

CHAPTER SEVEN

CENTER STAGE FOR THE HERMIT

One reason we call the missing person a victim is because of the involuntary situation of many victims' plight.

    —*Dennis Kelly,* Mountain Search for the Lost Victim

One day I'd like to hike the entire John Muir Trail and not leave a single footprint.

    —*Randy Morgenson, Charlotte Lake, 1985*

AFTER CHIEF RANGER DEBBIE BIRD'S phone call in 1996, Judi dialed Stuart Scofield without hesitation. She considered Scofield a close friend whom she associated with good times in Susanville, a time when Randy was passionate about photography.

Scofield had often joined them for dinner at their home, and inevitably he and Randy would migrate into Randy's office/darkroom to look over photos and discuss for hours the art of printmaking or an upcoming photography workshop they would be teaching together.

Since they'd moved to Sedona, Judi hadn't really stayed in touch with Scofield, but she knew Randy had. He'd stop by Scofield's house near Mono Lake before, and sometimes after, his seasons in the backcountry. If there was one person in the world who was Randy's unconditional friend and confidant, it was Stuart Scofield.

When Judi called, she was lucky to catch him at home. He was between photo workshops, working late at his desk, organizing and planning his summer schedule just as he had years earlier when he and Randy had taught outdoor photography workshops as a team. The workshops were unique: not only did they take place in the wild, but the students also lived in the wild during the workshops, car camping in national parks, with lectures around a campfire and morning coffee brewed on a Coleman four-burner stove. Randy's schedule dictated that he was usually available only at the end of the season—late September, and even that was iffy. Scofield was always worried that a search-and-rescue operation or some emergency would keep Randy in the backcountry, leaving Scofield to explain the absence of an instructor the students had already paid for. That had never happened. Usually Randy would make a dramatic entrance into the workshop camp: the rugged, bearded ranger coming in from the mountains, calves bulging, legs sinewy and strong, badge glinting in the sun.

Scofield loved making the introduction: "Everybody—this is your other instructor, backcountry ranger Randy Morgenson."

"Sometimes people would clap," says Scofield. "Those were just great days. Randy was so into photography, and into sharing his mountains."

When he picked up the phone and heard Judi's voice, Scofield knew something was wrong even before she asked if he'd heard from Randy. When she informed him that there was a search in progress, that he was missing, Scofield was immediately tongue-tied. He had been worried about his friend for some time.

But not wanting to jump to conclusions, he didn't reveal his true fears to Judi at the time. He just told her what she already knew—that Randy was having a hard time of it and figuring things out.

They ended their conversation with Scofield promising Judi that he would call if he heard from Randy and Judi thanking Scofield for being such a good friend to Randy.

◖

IT WAS THURSDAY, JULY 25, 1996, hours before the American flag would be hoisted up the mast at park headquarters. At Bench Lake, Rick Sanger was sitting upright, his naked legs still in his sleeping bag, a down jacket warming his torso, packing his backpack by headlamp and willing the light blue hue of morning to rise above the craggy peaks and banish the blackness of what had been a long, anxious night.

Officially deemed Operational Period I, this marked the first day of the search, the fifth day since Randy had gone missing. Four highly skilled two-person search teams would be inserted by helicopter atop 12,000-foot mountain passes, at the base of brush- and willow-clogged ravines, and amid the glacial rubble of eroding granite amphitheaters far from maintained trails. Remote, wild, beautiful country. All of the segments being searched had routes that converged eventually with Lake Basin, the center of a spiderweb of theoretical patrols that Randy might have taken.

George Durkee awoke early to the song of the hermit thrush—perhaps the same bird that had serenaded Randy five days earlier and inspired the final entry in his station's logbook: "Hermit Thrush sang briefly at the Bedouin Camp this morning."

The entry—one of the shortest of Randy's career—appeared hauntingly incomplete. Durkee remembered reading more detailed rhapsodies of Randy's favorite alpine songbird, like the one written when he was stationed along the shores of Little Five Lakes in 1977.

"Twilight," wrote Randy in his logbook, "up the slope and in the lodgepoles he prefers, a Hermit Thrush gives voice to a few tentative notes, then full song. High, slow, rising, crystalline notes. Drawn out, held, savored, then . . . descending into the fullness of sound which comes from deep in the chest.

"Quiet.

"Then he begins again. From high in the lungs and rising almost beyond the range of hearing. Clear, flutelike, ethereal.

"One of the world's mystical sounds. Ranking with the bugle of elk in frosty autumn forests, and the quavering laugh of the loon off misty, rain-drenched northern bays and lakes.

"If early people with mythology in their breasts . . . had lived with this undistinguished bird . . . perhaps the hermit thrush would have inspired legends and lore as did the loon.

"Twilight.

"Quiet.

"And the Hermit thrush singing. The world is poised, listening, alert.

"Center stage for the Hermit."

It seemed that Randy had traded places with the thrush—there wasn't a ranger in the park whose heart and mind weren't poised, alert, and focused on him.

From his sleeping bag Durkee could see that Sanger and Incident Commander Randy Coffman were already awake, milling about and sipping from steaming cups near the station. Rangers Lo Lyness and Sandy Graban were nowhere in sight, making this an opportune moment for Durkee to approach Coffman in private. He needed to confess that he hadn't been entirely forthcoming the night before while planning the search.

As he walked over to Coffman, Durkee could feel Sanger's energy from 50 feet away but he wasn't in the mood to talk, so he passed him by with barely a nod. Shoulder to shoulder, Durkee quietly spoke, "You should probably know a couple of things." He told Coffman that Randy had come into the high country with divorce papers, that he had seemed pretty depressed, and that Randy and Lyness had had an affair that was now over. This, explained Durkee, *could* mean that Randy *might* have left the mountains. He made it clear that though the possibility was low on his list, it was still a possibility. Mainly, he wanted Coffman to understand the dynamic between Lyness and Randy. He was still too conflicted to verbalize suicide, but left it at "Randy definitely hasn't been himself."

Coffman thanked Durkee for being candid and suggested he keep an eye on Lyness while they searched. Some of the terrain they'd be traversing required complete attention. Loose rock, steep terrain, and exposed cliffs—the last thing they needed was a secondary incident.

Coffman, who had already been radioed news of the pending divorce by Chief Ranger Bird, now had more of the picture. According to Durkee, Coffman casually approached Lyness before the helicopter arrived to make sure that she, being "close" to Randy, was okay taking part in this search. The message received in return was that she would be extremely upset if she were not taking part in it.

With that reassurance, Coffman filed away the information and dropped the issue.

BEFORE MARJORIE LAKE BASIN lost its early morning shadow, the parks' helicopter arrived to pick up Durkee and Lyness, who, as Team 1, had been assigned the highest-probability area of Lake Basin.

The pilot briefly skimmed the John Muir Trail north of the Bench Lake station and then gained altitude in a northwest arc, the left windows tilting down toward the wooded shoreline of a glassy Bench Lake. As the helicopter leveled off, the 12,900-foot polished walls of Mount Ruskin filled the right-hand window. A few minutes later, the helicopter set down in the broken granite atop the wide, dusty saddle of Cartridge Pass.

Seconds later, Durkee and Lyness were alone. They spread out to scour the sand traps and gravelly spots that were likely to hold a footprint or the telltale indentation of the ski pole they assumed Randy had been using as a hiking stick.

After more than an hour, they'd scrutinized this wide mountain passage and dropped down onto the north-facing slope that spilled into the basin. Snow was still in large patches, making for a slippery slog through loose talus chips and blocks that at times teetered when stepped on. Not far over the crest of the saddle, faded switchbacks zigzagged down the slope in barely perceptible lines, the remnants of an old sheep trail that was the original John Muir Trail before it was re-routed up Palisade Creek and over Mather Pass in 1938. Such was the history of the place Randy had shared with both Durkee and Lyness and, in happier times, Judi, whom he had taken this way on patrol to Marion Lake in the late 1980s. As Durkee and Lyness scanned the strips of soil held in place by the ancient switchbacks, Durkee remem-

bered Randy telling him about the rocky cliffs to the west, toward
Marion Lake, where Ansel Adams had propped his tripod for a precarious angle of the basin below. Randy had also recounted to Durkee
the "crime" the Sierra Club had committed during the 1920s, when
they'd brought one hundred mules into this basin for one trip. "How
could they?" Randy had said. "You wouldn't march one hundred mules
into the Sistine Chapel, would you?"

Now the trail, etched amid the talus, was long forgotten except to
those with an interest in history, a sense of adventure for cross-country
routes, or like Durkee and Lyness, for those searching for a missing
friend who favored these least-traveled areas.

"George, look at this," said Lyness. In a moist patch of earth near the
seeping edge of a melting snow patch, was a boot print, around a size 9,
Randy's size, pressed firmly into the soil. It was heading down into the
basin. Careful not to disturb the track, Durkee sketched the outline of the
sole on a piece of paper and marked the spot with a small cairn of rocks.

Durkee radioed to Coffman, "Without a search dog, there's no way
of telling whose track this is."

By now it was late morning, and as he looked out over the stunning
blue-on-gray mirage that was Lake Basin, Durkee's sixth sense as a
ranger kicked in. Everything in his being told him nobody was there.
He shook his head, almost as if he'd been duped. Doubt, he knew, was
a big part of the search-and-rescue mind game. He couldn't shake the
suspicion that he, all of them, had missed something while formulat-
ing the consensus the night before. Perhaps they had relied too heavily
on Randy's past affinity for Lake Basin.

After a sojourn in the basin in 1979, Randy wrote his parents and
told them, "I could be a backcountry ranger forever."

Ten years later, he lobbied to protect its fragile environment. In his
1989 end-of-season report, he described Lake Basin as being the only
large, classic, alpine and subalpine lake basin that was flourishing—
because of little or no use. At the time, there was talk of reestablishing
the Cartridge Pass Trail, and Randy had voiced his disapproval: "If
we do anything to increase human use there we will have destroyed

something we can't replace. I believe the backcountry management plan should state unequivocally that the trail between the two forks of the Kings River will never be touched, and that all stock be prohibited. . . . There is not another place in the park like this."

More recently, on August 27, 1995, he had described the basin simply as "one of my places. I feel I could spend my life here."

Still on the radio, Durkee told Coffman, "Something's not right here. We need to amp this thing up."

Having anticipated this, Coffman already had dog teams en route to the park. A dog and handler would be flown to their location at the northern base of Cartridge Pass the following morning. A well-trained scent dog could confirm or dispel Durkee's hunch that Randy was not in Lake Basin.

But had he at least passed through?

The two rangers spent the day circumnavigating the basin, checking known campsites, and another classic Randy route to Upper Basin via Vennacher Col. No other clues surfaced until late in the afternoon as they circled back on the far northern side of the basin, traversing the gently sloping higher ground that eventually steepened into two peaks. Between these peaks was another pass leading to Lower Dumbbell Lakes. In the gravelly soil approaching this pass was another track, though not as intact as the first. Again, there were no adjacent holes from a ski pole, but that didn't mean the tracks weren't Randy's. Hikers often carry a pole for a few yards or sometimes strap it to their pack, using it only for knee-pounding downhills. Sketching the sole size and building another marker cairn completed day one of the search for Lyness and Durkee.

By the end of Operational Period I, the four highest-probability areas had been searched on the ground. Graban and Sanger had searched the second-most-probable area: Marion Lake and the surrounding cirque. While circling the lake, they noted large, illegal fire pits near its shores. Knowing Randy's disdain for such resource-damaging eyesores, they deduced that he couldn't have walked past these without destroying them.

No clues were found in Segment B (Dumbbell Lakes, 2,894 acres), which was searched by rangers Ned Aldrich and Dave Pettebone. The team of Bob Kenan, who had taken time off from working on the park scientist's blister rust survey, and Dario Malengo, a ranger stationed in the Sequoia backcountry, located a set of prints heading north toward the top of the col between Dumbbell and Amphitheater lakes in Segment C, sans an adjacent ski-pole mark. Again, without a scent dog, the tracks couldn't be connected to Randy.

All four teams prepared to spend the night in the field.

Meanwhile, a mountain of paperwork was being organized to track the SAR on a daily basis. Incident action plans, debriefing forms, incident objective forms, safety memos, air operations summary sheets, and medical plans (for Randy or injured searchers) were just a few of the documents flying through photocopiers and printers that night at the Cedar Grove fire station.

The search teams were debriefed. Their estimated probability of detection (POD) for the segments searched that day took into consideration five potential scenarios: Randy was (1) mobile in the segment; (2) immobile in the segment; (3) responsive in the segment; (4) unresponsive in the segment; and (5) in the segment previously, but not currently. All the teams estimated that there was approximately a 50 to 60 percent likelihood that Randy would have been detected in the segments they had searched had he been mobile and responsive (the groups had called out for him periodically). But assessments dropped to 20 percent and less if he was presumed immobile and unresponsive, thus seriously injured. The terrain was such that it was impossible to cover every square inch. "A boulder could hide a person," says Sanger, "and there are a lot of boulders out there. You can't walk around all of them—you just take the likely route and keep your eyes and ears open." If Randy were seriously injured or unconscious, it was at least 80 percent likely the searchers would *not* detect him.

Back at the Cedar Grove incident command post, the overhead team reviewed the day's results and Coffman decided to pass off his incident commander position to Dave Ashe. Coffman named himself the operations section chief and gave experienced frontcountry ranger Scott Wanek the duties of planning section chief.

But Ashe would later call these on-paper designations "smoke and mirrors." Coffman was still running the show.

BEFORE DAWN BROKE on Friday, July 26, six days since Randy's last contact and the second day of the search, twenty-nine people, three helicopters, and one scent-specific tracking dog had already been assigned their duties for Operational Period II. At 7 A.M. the helicopters, including two from the military, began ferrying personnel thoughout the search area. In addition, a law enforcement team was mobilized to investigate the disappearance from a different perspective. Foul play couldn't be discounted.

Enter NPS Special Agent Al DeLaCruz and his two assistant investigators, Ned Kelleher and Paige Ritterbusch—all three commissioned law enforcement rangers. DeLaCruz, a 49-year-old Vietnam veteran with twenty-three years of service in the NPS, was the most senior law enforcement officer in Sequoia and Kings Canyon. It was his duty to teach the yearly refresher courses required for rangers to maintain their law enforcement commissions.

DeLaCruz immediately remembered Randy from training because his big bushy beard made him stand out from the other backcountry rangers in the classrooms. Randy also had quietly and respectfully approached DeLaCruz during defensive tactics baton training to tell him that maybe the backcountry rangers didn't need to be there because "we don't carry batons in the backcountry."

In truth, Randy and company perplexed DeLaCruz. For a dozen years, he had trained rangers in law enforcement tactics at three Southwest national parks and national monuments and he had "never seen anything like" the backcountry rangers at Sequoia and Kings Canyon. He didn't know it, but the rangers hadn't encountered anybody quite like him either—a by-the-book NPS officer who looked completely at ease in camouflage and on the firing range.

Sensing the rangers' resistance to law enforcement training, DeLaCruz had let them know that he had started as a seasonal campground ranger and trail-crew worker himself two decades earlier. "Listen, you guys," he said, "this is different for me. This type of training is in your best interest, and I'm not quite sure how to explain it. It's my philosophy that if you're going to be a law enforcement ranger, you should do it to the best of your ability. You're carrying a weapon and a badge, and that brings with it inherent hazards. Look, shit can happen, even out there where you're stationed."

Most of the rangers were receptive, and Randy himself had given a half-nod, half-shrug—a sort of "fair enough" gesture. But another backcountry ranger told DeLaCruz privately, "The only reason I'm carrying this gun and wearing this badge is so I can keep my job."

The frontcountry rangers were generally interested in the defensive tactics, disarming an armed suspect, controlling crazed individuals, and so on—likely because it was very possible that they would encounter such situations. The backcountry crew—the veterans, anyway—seemed to prefer nonlethal, nonphysical ways of dealing with aggressors. These tactics were great first lines of defense but not enough, as far as DeLaCruz was concerned. He knew that rangers—and not just those in the frontcountry—were statistically the most assaulted federal officers in the nation.

In the summer of 1984, a noncommissioned ranger in the backcountry of Sequoia and Kings Canyon was strangled into unconciousness after approaching a camper who had built an illegal campfire. The ranger survived and the suspect was apprehended the following morning by armed rangers, but the fact remained: "Shit happens."

DeLaCruz's main job was investigating crimes in the parks, apprehending poachers, and directing special undercover units that investigated various criminal elements, including drug gangs who were using the parks' lower elevations to cultivate massive marijuana gardens. Over the course of his career, he'd thought he had seen it all—ritual crimes, hallucinating suspects pursued by aliens—but he'd never had a missing backcountry ranger case.

At Randy's station, it proved to be even a little "creepy," because absolutely everything seemed in order.

DeLaCruz treated the Bench Lake ranger station like a crime scene. Not the dusting-for-fingerprints, yellow-ribbon-police-line kind of crime scene. There was no visible evidence—no sign of a struggle, no blood—to call for that level of scrutiny, but the station and surrounding area were thoroughly searched for clues. "Randy was a missing, potentially injured ranger, but he was also an unaccounted-for federally commissioned law enforcement officer. We had to assume anything and everything," says DeLaCruz.

He had flown into the mountains with a snapshot of Randy's life and current situation, conveyed by Chief Ranger Bird. He was aware that Randy had brought divorce papers into the backcountry, that he'd had an affair, that he might have been depressed. He also knew that Randy was probably the most adept ranger in the parks, well versed in search and rescue, and despite his gentle demeanor, he'd exhibited to DeLaCruz some proficiency in self-defense.

He began by opening the footlocker. Inside was Randy's duty weapon, as well as two personal diaries. After reading just a few lines, DeLaCruz could tell that Randy was unhappy. Figuring the diaries might contain some bit of information to help the search, he temporarily turned them over to Coffman. He wanted to send them to a profiler with the Department of Justice as soon as possible; it wasn't his place, or area of expertise, to interpret the writings of a man he didn't know. The divorce papers were nowhere to be found.

He mentally worked up several scenarios. The most probable theory was that Randy was injured somewhere, which didn't really concern him; there was already a boatload of people taking care of that possibility. Or Randy could have left the mountains. There was foul play to consider: without his gun he was at a disadvantage if he met a violent individual while on patrol. Of course, there was also the possibility of suicide, and DeLaCruz couldn't discount the idea that Randy was hiding out, even evading searchers.

In the frontcountry, the first order of business for the investigative team was tracking down backpackers who had passed through the search area in the previous six days. Kelleher was assigned the tedious job of wading through wilderness permits. Ritterbusch helped facilitate the posting of Overdue Hiker flyers at all the trailheads leading into and out of the mountains, and entered Randy into the California Law Enforcement Telecommunications System (CLETS) and National Crime Information Center (NCIC) as a missing person. Bulletins were sent to numerous law enforcement agencies, local hospitals, train stations, and bus depots.

In the backcountry, DeLaCruz scribbled notes on an ever-present pad of paper. Where was Randy's vehicle? Had his credit cards been used recently? Bank accounts accessed? His wife—he had to speak with Randy's wife. And interview all the backcountry rangers posthaste. And follow up with Coffman and the chief ranger. Who was the last to speak with Randy? See him in person? Who were Randy's closest friends in the parks? Outside the parks? Did Randy have any enemies? Had he experienced any confrontations? Medical issues? History of drugs or alcohol?

Something told DeLaCruz this incident wasn't going to be resolved anytime soon.

AN INDEPENDENT DOG HANDLER named Pat Bardone from Tulare County was flown to Cartridge Pass around 9 A.M. on the second day of the search. Cowboy, a bloodhound, was a scent-specific tracking dog—meaning he was trained to follow the scent of an individual, usually facilitated by a recently worn item of clothing. Durkee had anticipated this and brought a pair of Randy's hiking socks in a plastic bag. Bardone gave Cowboy a good sniff, took his harness off, and put him into "search" or, as Bardone put it, "scout" mode.

Nose to the ground, Cowboy trotted down the difficult terrain of the northern slope of the pass—right past the first track Lyness and Durkee had marked the day before. About a quarter of the way into the basin—more than a mile past the track and near the outlet of the first big lake—Cowboy "alerted," jumped up with his paws on Bardone's chest. "A good sign," according to Bardone's report of the day's events. With his harness back on, however, "Cowboy kind of bird-dogged all over the place," says Durkee, "not really seeming to follow any one track."

But then Cowboy moved deliberately toward Vennacher Col, where a section of cliff provided a good view of the basin, "the perfect vantage point for a photograph," Durkee had surmised the day before. Perhaps Randy had camped near the lake, waited for the right light, then slipped while climbing up the cliff. The area below was brushy and clogged with willows. Near the willows the dog seemed to be responding to scent. The team thoroughly thrashed through the thick foliage, knowing that if Randy was there he probably wouldn't be alive. Relieved to find nothing, they moved down the drainage, following Cowboy.

Lyness split off and took a different but parallel route, looking for "sign." "Lo figured Cowboy was just being a dog and cheerfully running around," says Durkee. "Neither of our confidence was really high." He stayed with Cowboy and Bardone, who continued down-canyon on a "dog route" that neither Durkee nor Lyness could imagine Randy would have taken, but Bardone was letting Cowboy do his job. Durkee began to lose patience, thinking, "There is no way Randy is out here in this flat area—he's not hiding under a bush." After some time, he urged Bardone to "encourage" Cowboy back up-canyon toward the pass leading to Dumbbell Lakes, the site of the second track they'd seen the day before.

Halfway up the pass, the dog alerted again and seemed hot on a trail. Just before the top, he stopped, sat down on his haunches, and stared straight ahead—"an enigmatic clue," thought Durkee, "that Randy had headed over to Dumbbell Lakes."

After a ten-hour day, Durkee, Lyness, and the dog team were flown back to Bench Lake, where debriefing forms were being filled out en masse by searchers. Under the section that asked searchers to estimate the probability of detection in their area, Durkee wrote, "Morgenson is <u>NOT</u> in Lake Basin/Area F (POD): 70%. Dog tracking existence of tracks that he <u>had</u> been there (POD): 40%." Lyness, on her own form, also estimated a 70 percent POD that Morgenson was not in the segment. She concluded with "The question remains unanswered as to whether or not he crossed the Basin."

On the same form, each searcher was also asked to describe "any search difficulties or gaps in coverage." Durkee voiced his (and Lyness's) skepticism regarding the search dog: "Hard to tell if dog was tracking scent; handler thinks high probability dog was on scent." Lyness wrote in her logbook that night, "Dog went in circles for 2–3 hours, not heading anywhere we thought Randy might go."

Meanwhile, word spread that backcountry ranger Dario Malengo had been medevaced out of the backcountry. He and Bob Kenan had been approaching Dumbbell Lakes from Amphitheater Lake, converging from the north on Durkee and Lyness's position south of Dumbbell Lakes. Unable to follow the usual trail due to snow cover, they'd improvised a new route that led to Upper Basin, essentially a cross-pattern across the route Randy may have taken if in fact those were his footprints leading toward Dumbbell Lakes. While descending a talus-clogged slot, Malengo "rolled" a loose chunk of granite. He reacted quickly and dodged the rock, but the hand placed for balance was crushed, resulting in a painfully broken finger.

At the only landing zone in the immediate vicinity, Amphitheater Lake, Malengo was whisked out of the mountains and Kenan abandoned the route to meet up with another team north of his location that was searching along Palisade Creek. As he picked his way down the canyon, following Cataract Creek to its confluence with Palisade Creek, he reflected on Malengo's injury, the snowfield, and Randy. The hazards were all around him: "a slip on that snowfield could have easily been fatal . . . nothing but sharp and jagged talus at the bottom."

In the late afternoon Kenan met up with the search team that, like him, had discovered no clues or tracks along these arterial backcountry routes. They radioed their position, and an hour later a military helicopter approached. They'd expected the smaller park helicopter, so the landing zone they had chosen was too tight for the large Huey.

After a long hover 50 feet off the deck, the pilot descended alongside a mound of granite, the giant whooping blades trimming the tips off a few pine trees in the process. Kenan watched someone he thought was a "hot-shot commando" jump out of the helicopter from a 4-foot hover and wave the searchers over. To his surprise, it was Jerry Torres, a veteran trail-crew supervisor, who had joined the search.

"That was when I realized this was going to turn into a massive search," says Kenan. "Usually the rangers handled the searches, with a few local volunteers. When personnel who traditionally handle other duties in the park join in, you know it's gearing up to be big."

Rick Sanger confirmed that he shared this same sentiment when he and Sandy Graban had returned to the Bench Lake ranger station on foot, after searching the south side of Cartridge Pass, down into the forested shores of Bench Lake and then back along the open benches and talus slopes forming the western wall of Marjorie Lake Basin.

In just one day, "Randy's quiet and pristine station had changed," says Sanger. "From a distance it looked like a High Sierra MASH unit, with tents pitched everywhere, helicopters coming and going.

"As I walked into camp, I was surrounded by the sounds of a search—chopper blades, people whistling and yelling in the woods, dogs barking—the sounds of trouble. It was very surreal, but then I had this funny thought of Randy walking in and looking around and yelling, 'What in the hell are you people doing to my camp?!' Kind of like 'You kids get out of my yard!' I just kind of quivered, keeping the totally inappropriate laugh in my chest as I claimed a spot that wasn't too close to everybody else."

While the initial search teams were reconvening after their two-day assignment, Task/Team 4—backcountry ranger Dave Gordon and Laurie Church, another trail-crew supervisor—were just beginning a two-day assignment. They would be spearheading the initial probes in the low-probability southern regions of the search area boundaries, near the headwaters of the White Fork drainage, with the intent of covering the area to Woods Creek.

Gordon would later express in his debrief that he and Church "were not given any info/briefing materials before search—that would have been nice to have. Randy was at Whitefork Camp w/ Laurie Church on July 11. Laurie asked if he wanted a re-supply of anything . . . (such as an extension of his stovepipe because his stove wasn't working well). Randy replied, 'No, it's not worth it to come back.' " Gordon also conveyed some potential routes that Randy had told Church he planned to check out.

The information from Church suggested that she was likely one of the last park employees to have spoken with Randy in person. Soon enough, she would be on DeLaCruz's list of people to interview.

DELACRUZ HAD SPENT THE DAY at the Bench Lake station, and with helicopters coming and going with search teams, he wondered if he wouldn't be staying the night as well.

Finally he was told that a helicopter was heading back to Cedar Grove in the frontcountry and there was a seat available for him. As the helicopter landed, George Durkee and Lo Lyness got out. There was only a small window of opportunity to ask a couple of questions, but it seemed that Durkee had the same intent.

Durkee approached the special agent and asked him if he had accessed Randy's footlocker. DeLaCruz affirmed that he had and Durkee followed with "Was his gun there?" DeLaCruz couldn't think of any reason why he shouldn't answer this question, so he did.

Outwardly, Durkee tried not to express any elation at the news; inside, he was thrilled. "Yeah, on cross-country patrols, sometimes he wouldn't carry it," he said.

Durkee had been obsessing about the gun for two days. He'd had visions of coming upon Randy's camp somewhere and . . . "You get the picture," he explained later. It was a relief to know that Randy didn't have the most obvious tool for suicide with him, which meant that maybe Durkee was being paranoid; maybe suicide hadn't been on Randy's mind at all.

Before stepping into the waiting chopper, DeLaCruz asked Durkee whether Randy had any enemies, anybody who might want to cause him harm.

Without losing a beat, Durkee told him that he'd had two altercations in the backcountry that had really shaken him up.

"When?" asked DeLaCruz.

"Just last season—you can read all about it in the station log at LeConte. Randy definitely felt threatened on two separate occasions. A climber and a packer."

Back at the Cedar Grove fire station, the overhead team was getting a clear picture of the day's events. In short, none of the tracks from the day before had led anywhere. The tracking dog had proved inconclusive. No substantial new clues had surfaced.

Most distressing, if Randy was still alive, he'd been out there, alone, for six full days.

Esther and Dana Morgenson
at their home in Escalon,
California, 1941.
(Morgenson family archives)

Baptized into the
world of camping in
Yosemite Valley—Randy
Morgenson gets a bucket
bath in water dipped from
the Merced River, 1943.
(Dana Morgenson)

Randy's childhood
backyard: Yosemite
Valley and Half Dome.
(Dana Morgenson)

Randy builds a make-believe
campfire in Yosemite Valley.
(Dana Morgenson)

Randy (back window) and Larry Morgenson (cowboy hat)
at play in Yosemite, late 1940s. (Morgenson family archives)

Randy and Judi's wedding at the Ahwahnee Meadow a few steps from the Morgenson home, November 22, 1975. (Morgenson family archives)

A park visitor took this rare shot of Randy in full uniform as he worked the Ash Mountain entrance station of Sequoia and Kings Canyon, June 1965. (Morgenson family archives)

Randy's summer home—Little Five Lakes ranger station— during the summers of 1977 and '78. (Bob Meadows)

Randy's beloved McClure Meadow, with the Hermit in the background. (Neal Larrabee)

McClure Meadow ranger station, Randy's summer home for seven of his twenty-eight seasons as a backcountry ranger. (Bob Meadows)

Randy in the backcountry, 1966. (Morgenson family archives)

Dana Morgenson on one of his popular Yosemite "camera walks," 1978. (Morgenson family archives)

A Back Country Management Plan

for

SEQUOIA and KINGS CANYON
National Parks

UNITED STATES DEPARTMENT OF THE INTERIOR
NATIONAL PARK SERVICE

Randy's backcountry
bible. (SEKI archives)

On patrol, Randy would usually leave behind his firearm—but he almost always carried his camera, capturing such images as these. (Gene Rose)

(Randy Morgenson)

(Randy Morgenson)

Randy and Judi ski-touring in the high country, 1978. (Bill Taylor)

Randy and Judi's winter ranger station in Tuolumne Meadow, 1978. (Bill Taylor)

Randy ice skating on Tenaya Lake, 1975. (Bill Taylor)

Alden Nash, Sierra Crest district ranger, 1975. (Nash family archives)

Mark Hoffman and Robin Ingraham atop Unicorn Peak, Yosemite, in November 1986, a year and a half before the accident on the Devil's Crags. (Robin Ingraham)

This is the last photo taken of Randy before his disappearance. He's the ghostly figure in the trees behind the trail crew (Laurie Church is at center with shovel) at the White Fork camp, 1996. (Rick Sanger)

A photocopy of Randy's last note, from the case incident records. (SEKI law enforcement archives)

June 21
Ranger on Patrol for
3-4 days   There is no
radio inside the tent — I carry
it with me.
    Please dont disturb my camp.
This is all I have for the summer.
I dont get resupplied
            Thanks!

Overdue Hiker

Randy Morgenson

54 Year old male
Height 5'8"
Weight 150 pounds
Longish black hair
Black and grey full beard
Brown eyes
Very tan

Randy Morgenson is a National Park Service Backcountry Ranger out of Bench Lake. He will be wearing a Park Service uniform.

Last known location: 7/21/96 at Bench Lake

If you have seen or contacted this person please contact a Park Ranger. If after hours contact Sequoia & Kings Canyon National Park Dispatch at (209) 565-3341

This flier was posted at all SEKI trailheads once the search-and-rescue operation began, 1996. (SEKI law enforcement archives)

The once-serene Bench Lake ranger station tent, transformed into the bustling backcountry incident command post. (Rick Sanger)

A photo of Randy's favorite high-country basin—Lake Basin—taken from Cartridge Pass where George Durkee and Lo Lyness discovered the first tracks during the search. (www.peterstekel.com)

Ranger George Durkee and Paige Meier at the Crabtree Meadow ranger station. (www.peterstekel.com)

Ranger Bob Kenan at the LeConte Canyon ranger station. (www.peterstekel.com)

Ranger Rick Sanger after a medevac near his Rae Lakes ranger station, 1997. (Sanger family archives)

Ranger Nina Weisman on the High Sierra Trail near Bearpaw Meadow. (Eric Blehm)

N351WM

Subdistrict Ranger Dave Ashe and Chief Ranger Debbie Bird on a snow survey in the SEKI backcountry shortly after Randy's disappearance. (Debbie Bird archives)

Alden Nash looks down into the lower third of Window Peak drainage. Note the late-season snow patches along the creek, the gorge at photo's center, and Window Peak Lake farther downstream, August 2003. (Eric Blehm)

Alden Nash compares a photo to the rocks at his feet, marking the exact "spot" where the radio was found. The low creek water leads to the rapids and waterfall just out of view downstream. This area was estimated to be under ten feet of snow or more during the search. (Eric Blehm)

Nevada National Guard helicopter pilots Darren Chrisman and Bob Bagnato in front of the OH-58 FLIR-equipped helicopter used during the search. (Erick Studenicka/Nevada National Guard)

CARDA member Linda Lowry with her search dog, Seeker. (Linda Lowry archives)

Looking back up toward Explorer Col from the approximate radio location. Subdistrict Ranger Scott Wanek is walking toward the gorge where, farther upstream, Seeker fell through the ice. This area appeared flat (filled in with snow) to Linda Lowry during the search. (Eric Blehm)

Radio

Randy at Tyndall Creek ranger station, 1988. (Gene Rose)

George Durkee, National Park Service Director Fran Mainella, and Judi Morgenson in Washington, D.C., May 2003. (George Durkee archives)

CHAPTER EIGHT

POLEMONIUM BLUES

. . . the blessings of one mountain day; whatever his fate, long life, short life, stormy or calm, he is rich forever.

—*John Muir,* My First Summer in the Sierra

Polemonium is a creature of the sky, the drifting clouds and the summit wind.

—*Dana Morgenson,* Yosemite Wildflower Trails

AN UNANNOUNCED HELICOPTER landing near a ranger station in the backcountry almost always means one thing. Bad news.

One of the most unpleasant duties for a backcountry ranger was delivering what were known as "death messages" to backpackers whose family back in civilization had been struck by tragedy. On September 20, 1980, it was Randy who got the news. Within a half hour, he had thrown some items into his pack, locked up his cabin, and was being flown out of the backcountry. Later that night, still numb from the news, Randy arrived at the hospital, where he hugged his mother and told her it seemed only yesterday that he and his father had been hiking together in the mountains. His father had been so vibrant, so alive. Randy had even written in his logbook the day they'd parted ways that the visit had been "A real treat. I hope when I'm in my 70s I'm climbing over Shepherd Pass, crossing snow, wading streams, eating cold meals, and sleeping on the ground."

Randy stood outside the hospital room, scared to enter—afraid that the last happy vision he had of his father strolling around the trail's bend would be forever overshadowed by one of him lying on his deathbed in a sterile hospital room far from the mountains he loved. "He did not want to see his father as he was now," wrote Esther in Dana's diary, "but wanted to keep fresh the memory of the wonderful week they had so recently had together." When Larry and Judi arrived, the family, along with friend Bill Taylor, consulted with the medical staff and a chaplain about their next move. "The doctor explained it all quite clearly," wrote Esther. "We hesitated. It was difficult but finally we asked that there be no further interference with the course of nature, that the outcome be left to God. We really knew he was no longer in this place except bodily. It was hard but we knew we must say, 'goodbye, dear one,' for the rest of our time on this earth."

Three days later, Randy, Judi, and Esther approached Yosemite from the east side of the Sierra, over Tioga Pass, and were greeted by the "biggest, golden orange harvest moon I have ever seen," wrote Esther, still recording the days in Dana's diary. "It was for Dana." The park was full of his ghost. There was a memory attached to every granite face, waterfall, and meadow. There wasn't a bend in the mighty Merced River that Dana hadn't photographed in all four seasons.

An envelope arrived shortly after their return to the valley. Inside, a tribute from Ansel Adams. It was to be published in a special Dana Clark Morgenson memorial issue of the valley's newsletter, the *Yosemite Sentinel.*

> *Virginia and I are deeply grieved by the loss of our dear friend, Dana Morgenson. He was indeed a vital part of the Yosemite experience, not only for his immediate friends but for the uncounted thousands of visitors to whom he revealed the beauty of Yosemite.*

Photographers and interpreters come and go, but few relate as closely as did Dana Morgenson to both the people and the Natural Scene. It is easy in this pompous age to scan the big things of the world and forget to see the small miracles of life around us; the morning light, the flowers, those intimate details of the world that combine to make it beautiful. Dana Morgenson achieved this for a multitude who probably might never have been aware of it all. He encouraged them not only to see but to make records of what they saw and felt with their cameras. His gentle persuasion opened doors of vision and comprehension.

Dana Morgenson will live long in the memories of all who knew him and shared his devotion to the infinite variety of nature and his dedication to reveal and protect it for our children and their children, for the ages of mankind on Earth to come.

The night before a heavily attended memorial service on September 27, the family dined at Yosemite Lodge in the Four Seasons Restaurant, famous for its walls covered with large-format black-and-whites by Ansel Adams. To their great surprise, Adams's "superb prints had been removed from the walls and replaced by Dana's beautiful color photographs."

Randy was awed by the gesture.

Soon after the memorial, Randy and Larry took their father's ashes to the summit of Mount Dana, where a brisk wind scattered them across the slopes of the mountain that had first enraptured Dana with the Sierra, and where the Polemonium grew—the flower that both brothers knew "leads others to heaven."

Upon his return to Tyndall Creek on September 30, Randy's mood was reflected in the brevity of the emotionless daily reports in his logbook: "Sept 30: Rock Creek–Crabtree; Clear and warm, 25 people; Oct 1 Crabtree, 3 people; Oct 2, Crabtree–Tyndall. Hot and smokey. No one." His personal diary told a different story: "Too busy for tears."

On the last day of the summer, Randy wrote a cryptic entry that would be understood only by those who knew the events of the season:

"Finis (sob!)"

When Randy returned from the mountains, he and Judi helped Esther settle into the home she and Dana had built in Sedona, Arizona. They asked if she wanted to live at their house in Susanville or move closer to them, but Esther was steadfast in her resolve. A new home elsewhere would have no memories of Dana attached to it. For now, memories were her lifeblood.

In Sedona, shortly after Dana's death, Esther received in the mail a box of copies of his final book, *Remembering Yosemite*, which showcased his photography. Its dedication read: "To Esther, who has shared these memories of thirty-five happy years and helped to make them happy." Just weeks before they'd left for Alaska, Dana had put the finishing touches to the quotes and finalized the photos—in effect, writing his own epitaph.

Esther never let on to Randy and Judi how hard it was for her alone in that house. She had a number of friends and, from the outside, it appeared that she plunged into her new life in the desert with vigor— painting, attending luncheons, bird watching, joining the Audubon Society. Randy and Judi considered this workaholic lifestyle a continuation of the active life she had led with Dana in Yosemite, and her way of healing.

But inside she was struggling. For months after the memorial, she'd pick up Dana's red Standard Diary for 1980 and continue where he had left off. Each entry was in the form of a letter to Dana. On October 26, 1980, she wrote, "My darling, my darling, can I talk to you now? I ache for you—to feel your warmth . . . to see your smile, the sparkle and the crinkle around your eyes, to hear your voice—such a lovely voice. How many times we have sat by this fire where I am now alone. How many times we have spoken of the future and our plans. We thought there would be so much.

"It is a lovely little house—this. But where has its heart gone? Where is it? Where are you? Oh where are you? I have such a need for you.

"The fire flickers and dies."

IN THE SPRING OF 1983, Randy attended the workshop "The Photographer and Wilderness" in Kanab, Utah. The weeklong course taught by Dave Bohn and Philip Hyde was part of a personal resurgence of interest in photography that Randy experienced after the death of his father. The course verified some of Randy's philosophies surrounding the ethics of wilderness photography and a photographer's relationship with the landscape.

Randy had for years been trying to understand the emotions of the wildlife he communed with. He tried to "sense" the permission of the subject before taking its picture. For him it was all about respect, similar to the Buddhist reverence for nature.

Dave Bohn shared a similar philosophy, as illustrated in his 1979 book, *Rambles Through an Alaskan Wild: Katmai and the Valley of the Smokes*. "I want to know if a tree—any tree—really wants to be photographed," he wrote. "I have asked numerous trees this question but am not yet clear on the answer. I like to think, however, that if the photographing is done with sufficient respect, privacy will not be invaded."

Randy left that workshop with a renewed hope for the future of wilderness and an excitement for the coming summer, when he would truly concentrate his efforts to photograph the landscape "respectfully." He realized the one bit of photographic advice he'd never gotten from his father or Ansel Adams was the need to nurture an emotional bond with his subject, something portrait photographers had always capitalized on. To incorporate this into landscape photography would be a little "out there" to the masses, but for Randy it was a slap on the forehead. It made perfect sense. Combining the sensitivity of Bohn's philosophy with the unsurpassed technical teachings of Ansel Adams was like the founding of a new religion.

With this revelation came a desire to spread the word.

But Randy's plans to respectfully photograph the natural world were sidetracked temporarily when—shortly after settling in at Tyndall Creek ranger station—he was assigned by his SEKI supervisor to take pictures of "steel birds."

Unauthorized low-flying military aircraft, particularly jets with "cowboy pilots" from the twelve Army, Navy, Air Force, and National Guard military bases within "striking distance" of Sequoia and Kings Canyon, were creating sonic booms that reportedly triggered rock-slides and invaded the serene wilderness experience.

On August 12, 1983, a SEKI helicopter barely avoided a midair collision with an F-106. Officials from the bases firmly denied that they overflew any national parks or monuments below 3,000 feet.

Randy, who had taken to carrying an aircraft identification book in his pack, begged to differ. For twelve days that season, he camped out with his camera at various vantage points known to attract hot-dogging pilots. He called it "steel bird–watching."

On November 8, 1983, he received a letter on United States Department of the Interior letterhead.

Dear Randy:

I want to personally thank you . . . for all the special effort you put into trying to identify the low flying military jets in the Kern River area last summer. Through your long and boring days of sitting and waiting for jets to fly by, you were able to accomplish the next to impossible job of obtaining clear photographs of the tail numbers. These photographs should be adequate proof to convince the responsible Base Commanders that the violations you have been reporting so often do in fact exist.

Thanks again for all your special efforts to help us resolve the critical situation in the Kern Canyon.

Sincerely yours,
Boyd Evison
Superintendent
Sequoia and Kings Canyon National Parks

RANDY MET STUART SCOFIELD in 1983 at Lassen College, where Judi was teaching ceramics. Scofield had gotten into photography as a means of documenting his big-wall climbing of the 1970s. Ten years younger than Randy, he had grown up in Big Oak Flat, just outside one of Yosemite's entrances. He and Randy hadn't met then, but Scofield had known Randy's father by reputation.

Their enthusiastic, mutual interest in photography quickly made them good friends. They came from the same school of thought, having studied *The Basic Photography Series* by Ansel Adams. Scofield was just beginning to make a humble living teaching photography workshops and, recognizing the intense creativity and excitement that was generated each time they spoke about the craft, invited Randy to instruct a class with him in 1984 via the Sequoia Natural History Association. Others followed over the years.

During these workshops, Randy tried to find ways to express the photographic process. In time, he settled on an "invisible" approach to photography—invisible in that he did not want his style to become recognizable because any hint of the photographer's presence only took away from the sublime beauty of the subject.

"If the photographer is primarily involved with his own opinions and feelings about his subject, the photograph will probably contain more of the photographer than of that photographed," Randy would tell his students. "We have all seen this in portraits. The photographer can work for something particular and essential about the person he is photographing, or he can use his subject to express some artsy or humanistic notions of his own.

"I believe that the same opportunities occur in landscape photography, and my choice is to leave myself out to the extent that I can, and hope that the land can speak directly through the photograph.

"Admittedly, I place my tripod in a particular place, and make a host of other decisions, but if I am receptive to the place I am photographing, rather than thinking about manipulating it into a proper composition, something about those rocks and trees may come through the photograph, apart from my notions of what makes a good picture.

"I prefer to be a witness, over an interpreter."

Some of the students got it, while others, no doubt, thought Randy had spent a bit too much time alone in the woods.

Scofield could relate to Randy's viewpoint. The two photographers would "Zen out" in their conversations about the mechanics and philosophy of photography. One time they saw each other, in passing, in the parking lot of a Susanville grocery store. It was raining, but no matter. During a previous conversation, they had been contemplating aspects of Ansel Adams's landmark zone system—a processing system Adams developed so both amateur and professional photographers could predict the varying shades of light and dark on darkroom-processed prints—and now they were compelled to finish that discussion.

The conversation evolved, as it always did, and the two were so engrossed that they were unaware of their surroundings. For two hours they stood in the pouring rain. "That was when we became cognizant of the little island of high ground where we were standing," says Scofield. The entire parking lot had flooded. In fact, they'd forgotten what they had come to the grocery store for in the first place. "We laughed long and hard at ourselves, but that's how it was. Randy and I would lose ourselves in discussions."

Over time, Scofield noticed that Randy's bond with the Sierra was at a level he had never encountered in anyone else and how curmudgeonly opinionated Randy was about *his* mountains. Among other things, "People who moved too quickly over the land were, in his opinion, disrespectful," says Scofield.

Randy called them "trail pounders," and he couldn't understand them. Once he met a fast-moving "lad" coming toward him on the stones that cross the outlet below Helen Lake. He recorded the encounter in his logbook:

"Stepping to the edge of a large rock, he motioned me by, obviously not noticing my shirt with the badge. But just as I passed, he spotted the NPS shoulder patch. 'Oh, oh! Uh, wait!'

"He had 'two quick questions.' 'How's the snow on Muir Pass?' 'It's just fine,' I replied cheerfully, but it went right past him. 'Okay, good. And, uh, do you know, what's the fastest time the Muir Trail's been done in?'

"I just laughed. Another Muir Trail marathoner. And this one is going to make a record, having already 'done' the PCT [Pacific Crest Trail] in what he calls record time of 110 days.

"What is this infatuation with 'est'? Why are we beating our brains on a hard surface to be fastest, biggest, richest, on and on ad infinitum ad nauseam? I asked how many Audubon's warblers he'd seen or hermit thrushes he'd heard and he grinned sheepishly, looking down at his bootlaces. But this was an unfair question. Such a hiker has probably never slowed enough to notice, but I continued: 'Have you tried meadow sitting or cloud watching?' 'Anyone can do that,' was his response. There it is again. Machismo. This fellow is going to achieve, be a first, do things not everyone does or even can do. That becomes his goal.

"We're a restless breed, we moderns. Hardest it is to sit still and be attentive to our surroundings. Boredom comes to most of us very quickly."

Alden Nash took note of Randy's sometimes condescending, if not self-righteous, tone in his logbooks, but he felt that if a backcountry ranger could vent on paper and remain a friendly and cordial wilderness host while interacting with the public, that was a fair trade-off.

"No doubt about it," says Nash, "Randy was opinionated, and if he was ever cynical or condescending in public, he did it in a way that flew right over most backpackers' heads. Bottom line, Randy was a good ranger—one of the best to ever wear the uniform."

Nash put his money where his mouth was, so to speak, on December 8, 1981, when he gave Randy an Outstanding Performance Award for a flawless record as a seasonal ranger, the first official award Randy had ever received.

"Dear Randy," wrote Nash. "This is to compliment you on your Outstanding Performance for 1981 and for the past few seasons. The National Parks and the park visitors have benefited from your work as a seasonal Park Ranger for the past 14 seasons. Your knowledge and experience in the Backcountry Ranger position is unsurpassed in these parks. Both seasonal and permanent Rangers look to you for information, ideas, and inspiration on the job."

Nash went on for a page and a half of single-spaced accolades, recounting Randy's accomplishments before ending the letter with "In short, your overall attention to detail, your perspective and personal priorities concerning the job, and your experience and job related skills add up to an outstanding work performance. It is my pleasure to present you with this award."

With the letter was a government check for $350—not chump change for a seasonal ranger.

DESPITE RANDY'S curmudgeonly stance on wilderness issues, he was considered almost exclusively a kind and gentle man and ranger who made a positive impact on hundreds of wilderness travelers, many of whom called upon his medic skills and calm, composed, and reassuring nature during times of crisis. The parks' superintendents and chief rangers over the course of Randy's career received dozens of letters and verbal commendations beginning his first season, 1965, and continuing till the year he disappeared.

One letter, left on Randy's cabin door in LeConte Canyon, was from a woman who apparently had hiked into the mountains for some emotional healing, but seemed conflicted about whether or not to stay. The wilderness was a scary place. Where to find reassurance?

Dear Randy:

I wanted you to know that even though I only briefly met you, you have stayed with me. A couple of times I thought of walking up to say hello but I knew I needed to be alone and go through the things I went through. Sometimes I thought I was crazy and wanted to leave, but you can't run away from yourself (and someplace in me didn't want to). Anyway, I finally began to feel calm and more accepting and was able to let more of that gentle meadow into my heart. . . . This place definitely tried to care for me and help me find a more caring place inside of me. I could see that God was all around me. . . . Anyway, I did finally feel more open and more at home here. Thanks for your help.

Nancy

To many, Randy personified the wilderness steward, not a wilderness cop, as exemplified by the following letter, written to the chief ranger in 1985. The backpacker, on an "enjoyable" hike from Onion Valley to Sixty Lakes Basin, had chosen a campsite he thought was above the No Camping area surrounding Bullfrog Lake:

> *The next morning, as we were packing to leave, Randy Morgenson, the Park Ranger at Charlotte Lake, came by and said that our camp had been in the area intended . . . to allow the area around the lake to recover. After hearing my description of the way we had selected the site, Mr. Morgenson issued us a Courtesy Tag as a reminder to avoid this mistake in the future. I want you to know how very courteous Randy Morgenson was in this situation. In all my 25 years of backpacking in the Sierra Nevada, and encountering a number of Forest Service and National Park Rangers, I don't recall meeting a more considerate person. I hope you can convey this thought to him in some appropriate way. Thank you for the assistance you provided our party specifically through Randy Morgenson and in general by preserving a beautiful natural area.*

Judi Morgenson, designated a Volunteer in the Park, was often a part of these accolades. In 1985, a man wrote to the superintendent:

> *The Morgensons were literally lifesavers. I suffered pulmonary-edema and had to be copter-lifted out of Charlotte.*

. . . Without their help, I might be history right now. They acted quite professionally as they performed their duties. I hope the park decides to keep these wonderful people. In an era of cutbacks, we can't afford to lose good rangers.

The letters continued. In 1986:

Exceptionally helpful. Told me what to expect ahead, where to cross streams, and where I could best camp. Polite, honest and willing to listen to me. I will remember him as an exceptional ranger willing to assist.

To the superintendent that same year:

I met your backcountry ranger Randy Morgenson on a recent trip over Bishop Pass. We experienced heavy snows on September 23 and 24 and found him to be helpful and accommodating. His devotion to the country and to his job was certainly a credit to your organization and I thought you should know about it. He also is a hell of a great maker of buckwheat pancakes.

In 1988, to the superintendent:

Your ranger, Randy Morgenson is to be commended for service to the public above and beyond the call of duty. Last year . . . my wife sustained an acute low back strain at upper LeConte Lake. After a two day layover, we decided to return over Bishop pass because of her painful infirmity.
Ranger Morgenson was kind enough to come up from LeConte Station and carry her pack down. The next day he came up again and carried her pack over the pass to a high lake.
This year we tried it again, but . . . I experienced an acute ulcer attack at Sapphire Lake which incapacitated me for a couple days. We saw no one for three days until we reached McClure

meadow. Ranger Morgenson . . . arranged the next morning for
horses to take us out from Piute Creek to North Lake. . . . I would
not have been able to climb out unassisted. Throughout these
trials, Ranger Morgenson was very encouraging and supportive,
which was a great comfort to both my wife and myself.

In 1986 a woman from Palo Cedro, California, slipped and fell and
was unable to bear weight.

The next day my husband discussed the situation with Mr.
Morgenson, who examined my ankle, made suggestions for
alleviating pain, and caring for the ankle in case of a possible
fracture, or torn ligaments. He showed deep concern for the
situation we were in.

The woman explained how Randy arranged for a helicopter flight
the following day, after her husband and son hiked out.

This meant I would be spending twenty seven hours in the
backcountry by myself. During this time Mr. Morgenson was
extremely kind. He showed utmost consideration of my situation,
checking that I was all right, bringing me fresh water, and when
he learned that I'd sent our stove with my family, he brought me
some hot food.

Mr. Morgenson said he was just "doing his job," but I am sure
he has many duties, and having an injured hiker on his hands
only complicated his job. Yet he never showed this in his manner,
and was always very patient and kind.

His wisdom and expertise on the proper care of this type
of injury has been proven, as subsequent examination by a
physician showed that I do indeed have torn ligaments. . . .

We want to thank you for employing such a fine person in
your service, and indeed making the LeConte Canyon area a
safer place to be.

More praise:

I'm writing to express my appreciation for and sing the praises of your ranger Randy Morgenson. Last Wednesday we were in dire need of help—one of our party was stricken with severe abdominal pain and needed medical attention. We were in the Forester Pass area and were fortunate to find Randy on the trail. He was superb! We were anxious and concerned, naturally, and he handled all of us with great concern and professionalism. He inspired great confidence and gave us all tremendous peace of mind, and had Mary out to help in no time. You have a truly excellent man on your staff and he deserves recognition and appreciation.

The public applause for Randy can be summed up in one sentence from a letter written by the parents of a boy who was airlifted out of the backcountry for a medical emergency: "It's great to know people like you are around when we need you."

Randy did seem to have an uncanny knack for being in the right place at the right time for wilderness travelers in need of assistance: Super Ranger, sans the phone booth and cape.

THE DEVIL'S CRAGS shoot up out of Kings Canyon like the crumbling, rotting teeth of their satanic namesake—jagged, constantly eroding remnants of black metamorphic rock that is ages older than the hard gray granite dominating the Sierra range. In 1988, Randy was stationed in LeConte Canyon, 8 miles from the Crags, the closest backcountry ranger station to these charcoal pinnacles. Randy considered the Crags geological wonders that were alive and constantly evolving. In early August, he was reminded that they were also one of those volatile areas in the Sierra poised for trouble.

Robin Ingraham Jr. and Mark Hoffman were two experienced climbers who lived in Merced, California, not far from the famed climbing walls of Yosemite. Though Hoffman was known in Merced's climbing community as "the mad soloist," the truth was that—until he met Ingraham in 1985—he just hadn't found a reliable partner whose climbing appetite and skills matched his own. For the next four summers, the two climbers sojourned into the High Sierra, bagging more than a hundred peaks. Not a day went by that they didn't either climb or talk, "a friendship," says Ingraham, "that comes once in a lifetime, if you're fortunate."

They also shared an interest in preserving mountaineering history. In 1988 they created a program reminiscent of the early Sierra Club's efforts revolving around summit registers: checking and replacing damaged peak registers on mountaintops across the Sierra. They collected the registers and delivered them to the Sierra Club archives.

On August 11, 24-year-old Ingraham and 28-year-old Hoffman awoke at 4:30 A.M. to begin their ascent of Crag Number 9, a peak first climbed by Jules Eichorn and Glen Dawson in 1933 via the class 4 right side of the northwest arête. Their goal fifty-five years later was the left side of the arête. Still class 4, but a route nobody had ever climbed.

Ingraham and Hoffman picked their way up the loose, crumbling rocks in the predawn hours. It was the twenty-second mountain they'd climbed that summer, but for some reason Ingraham felt an unprecedented anxiety. Hoffman, noticing his partner's unusually slow pace, asked, "You all right?"

"I'm not doing this," Ingraham answered with conviction. "I've got a weird head today," he added.

Hoffman pulled a rope from his pack and said, "Let's be safe and rope up. I'll lead."

Perhaps the anxiety arose from a moment when the two had been traversing the west face of Crag Number 5 two days before. Ingraham had reached a large ledge and leaned against an "automobile-sized" boulder that shot down the cliff. "Despite our climbing abilities," says Ingraham, "they [Crags 5 and 6] tested every move. One hold after the other seemed to break with the smallest body weight." With "the greatest care," they had reached the summit and found and replaced the fragile Crag Number 5 peak register that Sierra climbing icons David Brower, Hervey Voge, and Norman Clyde had placed there in 1934.

After roping up on Crag Number 9, Ingraham's uneasiness subsided. Finding no summit register at the top of their first ascent route, they built a rock cairn and left a small book inside one of the weatherproof PVC canisters they'd brought along for that purpose.

Just past noon, Hoffman mused from the summit, "There's Crag Number 8. We should bag it while we're up here. Plenty of daylight." Ingraham hesitated for a moment, then agreed. They were in prime shape and could quickly rappel from Crag Number 9 to the saddle, climb the less technical Crag Number 8, and be back in their camp near Rambaud Lakes before dark.

True to form, they were atop Crag Number 8 at 3:30, marveling at a spectacular group of clouds forming around Mount Woodworth. They felt as if they were in an Ansel Adams photograph—black mountains and white clouds, the only color in the landscape a surreal blue sky that darkened with thunderstorms to the north before their eyes. Time to get off the peak.

Sure-footed yet cautious, they descended the west side of the cirque and entered a class 2 gully that would take them back to camp. About 300 feet from the bottom of the gully, it steepened into a chute filled with loose rock that split into a fork. Centuries of rockslides and snow avalanches had run between these walls, leaving an obstacle course of loose debris to slip, slide, and negotiate.

A few yards above the split, a refrigerator-sized boulder sat perched, seemingly solidly in place. Hoffman tested it for stability, then traversed beneath. That was when it shifted, bringing the entire chasm to life in a violent rockslide that swept Hoffman off his feet and pulled him down the left-hand fork of the gully. The roar was deafening. Ingraham, who had been standing on solid rock only a couple of steps above where the slide began, watched in horror as his partner—unable to self-arrest—rocketed down the steep incline out of view. Hoffman came back into view 40 or 50 yards away, just as an airborne rock the size of a bowling ball struck his head. Then he disappeared into a void.

Charged with adrenaline and fear for his friend's life, Ingraham ran down the opposite gully. He found Hoffman lying amid the jagged rubble of talus at the base of the 50-foot cliff over which he'd fallen. Fearing the worst, Ingraham yelled. Hoffman sat up and Ingraham heaved a sigh of relief.

But that euphoria was short-lived. Hoffman collapsed when he tried to stand. "My leg is broken," he shouted in pain.

A cursory exam revealed that Hoffman had also broken his arm and suffered a serious head laceration. Internal injuries, if any, were unknown. Hoffman screamed as Ingraham reluctantly honored his request, straightening his grotesquely broken leg and building a rock cradle as a brace. All the while, Ingraham's mind was working, realizing that they were far, far from help. Then he remembered the ranger's cabin they'd passed in LeConte Canyon.

Ingraham carefully dressed Hoffman in all of their warm clothing and announced that he had to get help. Hoffman begged him to stay. "Please don't leave me. Please don't leave me," he said over and over again. But Ingraham knew that his friend was seriously injured and probably bleeding internally.

Part of the draw for coming to this location had been its remoteness. "Seldom traveled" was an overstatement. Entire seasons passed without anybody attempting even the approach to Ingraham and Hoffman's base camp, which had begun with an icy wade across the Kings River just south of Grouse Meadows, then headed into a horrendously steep bushwhack through waist-deep, skin-tearing manzanita topped out by a jigsaw puzzle of increasingly steeper granite slabs. Nearby assistance was a zero possibility.

The only chance for survival this far into the backcountry rested on Ingraham's physical fitness.

At 4:30 P.M., Ingraham told his partner, "You'll be okay. I'm going to get a chopper. I'll be back soon." He left behind a water bottle and at Hoffman's request gave him a bottle of prescription Tylenol with codeine.

The landscape became a complete blur for Ingraham as he ran a talus-strewn cross-country route toward their camp. At 5 P.M., rain and then snow began to fall. Thunder growled and lightning flashed in a darkening sky. "All I did was run and pray," says Ingraham. "I prayed to God for mercy."

At 6 P.M. he rushed into their camp and grabbed a flashlight, batteries, candy bars, and dry shirt. He clutched briefly at his sleeping bag but tossed it aside: "Too much weight."

After spending little more than a minute grabbing gear, Ingraham was off and running toward a path that would take him to LeConte Canyon and the trail intersection to Bishop Pass, which was adjacent to the ranger station. From the site of the accident, the cabin was more than 10 miles away. If nobody was at the cabin, a long, grim uphill hike awaited him; it was another 15 miles to their car at the trailhead. Twenty-five miles at altitude after a day of climbing that had already worked him to the point of shaky legs and burning muscles.

As Ingraham and Hoffman were climbing Crags 8 and 9, Randy was on patrol to Echo Col, some 8 miles north of his LeConte duty station. Around the time of the rockslide, Randy was caught by storm clouds, probably the same ones Ingraham and Hoffman had seen gathering to the north. Of his descent off Echo Col, Randy wrote, "Graupel showers on the way down," which made the route slippery and wet. Up high, the ground was white, making it look like winter. It would be a cold night in the high country.

Randy generally enjoyed leisurely headlamp or moonlit nighttime strolls back to his cabin at the end of patrols. But on this day, the gathering storm quickened his pace and he made it back to his cabin in record time, just after nightfall at around 8:15 P.M. Crossing the rushing creek in the conifer grove near his cabin, he saw a glimmer of light through one of its windows.

Cautiously, he walked around the front of the cabin and found the door shattered and a lone figure leaning over his table with a flashlight in his mouth, illuminating something. The cabin had been locked; a note on the door had said Randy would return that afternoon, and he was more than 15 miles from the nearest trailhead. Whoever this person was, he clearly didn't have permission to be inside his home.

"Hey!" yelled Randy. "Why are you inside my cabin?!"

The response he heard was "Thank God you're here!"

"That," retorted Randy, "doesn't answer my question."

At that moment, Robin Ingraham sank to the floor. "My friend is hurt. You've got to get a helicopter here quick. He might be dying."

The young man's frantic tone was genuine. "Slow down," Randy said, helping him to a chair. "Tell me what happened." Ingraham recounted the accident while Randy calmly lit a lantern and picked up a notepad and his radio: "Dispatch, this is 113—please close all park channels, we have a SAR in progress."

"We need a helicopter now!" Ingraham broke in frantically.

Randy gently pointed his palm at Ingraham to stop. He set down the radio and focused his gaze on the floor. Then he looked up and said, "Robin, we need to think this through. I need to coordinate the rescue with Fallon Naval Air Station. Maybe we should hike back to him right now. There are no air evacuations in the mountains at night. The Sierra is too high and rough. It's too dangerous. Those helicopter rescues are television fiction or military operations. What do you want to do?"

"He probably won't last the night," Ingraham choked out between sobs. "But. There's. No way we can get to him on foot in the dark."

"Robin," said Randy, "I'm sorry, but all we can do is wait until dawn. Let me coordinate the rescue."

Over the next hour, Ingraham listened as Randy organized logistics with his supervisor, Alden Nash. Alternately, he argued with the dispatcher—his relay to someone at Fallon Naval Air Station who said the station's airships, which, unlike the NPS helicopters, were equipped with a rescue hoist, weren't available. Between radio calls, Randy provided Ingraham with warm clothes and dinner. "Looking back, Randy was so kind to me," says Ingraham, "and at the same time he conveyed such confidence. I tried to sleep per Randy's directions sometime around midnight, but I just lay there till he lit the lantern again at 4 A.M. My mind was with Mark."

Randy tried to get Ingraham to eat a big breakfast, "as one never knows how long these days will be," wrote Randy in his logbook. Then the two hiked to a nearby meadow that was covered with frost—not a good sign. Ingraham's heart sank. He started praying again.

Dawn broke, and still the air was silent. No thumping helicopter blades. Randy patted Ingraham's back and walked a few feet away, where he paced back and forth with the radio to his mouth—noticeably disgruntled at the tardy Park Service helicopter. An hour after sunrise, the helicopter finally approached from down-canyon, and a few minutes later, they were lifted skyward along with two Park Service medics.

The pilot indicated that the terrain on the southwest side of the Crags—where Hoffman was—was too steep and treacherous to attempt a landing, so he ascended the ridge and was able to insert the three-man rescue team plus Ingraham on a giant granite slab on the northeast side. Retracing his route from the evening before, Ingraham couldn't believe he'd run down the gully without falling. Then he realized how fortunate it was that Randy had shown up at his cabin when he had. Five minutes later, and Ingraham would have been off to Bishop Pass, some 15 miles from the ranger station. With the temperatures dropping, Ingraham, who had been physically exhausted, wet, and cold, was certain he would not have made it. Foremost in his mind was that he hadn't been taken out by the slide. He let himself grow optimistic that his friend would be alive.

As they reached the base of the gully, they could see Hoffman's brightly colored clothing beneath a 50-foot cliff.

"He fell off *that*?" Randy asked.

"Yes," replied Ingraham.

"That doesn't look good," said Randy, who let out a loud whistle and yelled, "Hey, Hoffman!"

There was no response.

Randy placed a hand on Ingraham's shoulder and said, "Robin, you're going to have to be real strong." Then he unshouldered his pack and walked up to Hoffman, kneeled, and shook his arm.

It was too late.

Ingraham sat down where he'd been standing. "My mind emptied as the weight of the mountains seemed to crush me," said Ingraham later. "I hated the mountains and regretted not sitting with Mark into the night so he wouldn't have died alone. I prayed that God allowed him the opportunity to be forgiven."

The coroner would report that in his opinion Hoffman had likely died, sixty to ninety minutes after Ingraham left for help, of "shock from injuries sustained." Those injuries included a fractured pelvis, left femur, and right arm; a separated back; a head injury; and internal injuries such as a ruptured spleen.

Since it was deemed too dangerous to evacuate the body up and over the Devil's Crags to the landing zone, Randy called upon Fallon Air Station, which this time sent a large military helicopter with a hoist. Ingraham watched his friend being hauled skyward in a body bag, a vision that would forever haunt him.

Randy helped the emotionally defeated climber pack up his camp and Hoffman's belongings. Then the park's helicopter delivered Ingraham to Cedar Grove, where Hoffman's body had been transported.

As he said goodbye to Ingraham, Randy was informed of another search-and-rescue operation in progress on the Hermit. It was a busy day in the mountains. He gave Ingraham another long squeeze on his shoulder and said, "I'm sorry." Randy's instructions were to stand by, so he did, hoping this SAR wouldn't also end in a Code 13. One body recovery in a season was bad enough. To have two on the same day would be unthinkable.

Patience was usually a state of mind Randy achieved easily. But this time his empathy toward Ingraham's plight made it impossible to relax. Hating the hurry-up-and-wait routine, he picked up his radio and contacted the wilderness office at Cedar Grove. He asked the dispatcher to make sure somebody watched over Ingraham. "He should not be left alone," he said before signing off.

Nina Weisman, who heard this call and volunteered to sit with Ingraham until his family arrived, was a second-year trailhead ranger, a recent college graduate at the threshold of a long career with the National Park Service. At the time, she dreamed of someday becoming a backcountry ranger and living alone in the far reaches of the parks. She knew Randy by reputation and considered him the ultimate ranger mentor. "I was impressed and touched by Randy's actions that day," says Weisman, "because he'd made the effort to follow up on the well-being of this young climber who he'd assisted, even while en route to another rescue."

To summit the Hermit, which Randy had stood atop numerous times, required a confusing scramble. At 12,352 feet, the Hermit's exposed granite dome has been battered by the elements for millennia. Climbers liken its cracked face to the skin of a weathered mountaineer, head thrust into the clouds and set slightly apart from its nearest granite neighbors. Randy loved the Hermit's distinct personality and the thunderheads that regularly gathered around its summit, grumbling like grouchy old men keeping the Hermit company.

Visible anywhere from the lower Evolution area, the final crux had proven problematic for solo climbers in the past, so Randy guessed this was where the fall had occurred.

He was wrong. This time, another experienced climber by the name of Douglas Mantle was the victim of another loose rock. He'd taken a serious fall near the summit crux and was unable to descend without assistance.

Anybody who's climbed a peak in the Sierra, then or now, would recognize Mantle's name; he has signed virtually every register on all the major and most other summits multiple times. The Hermit in 1988 would have been the then-38-year-old climber's 199th peak during his attempt to complete the Sierra Peak Section's "list" (247 mountains) for the third time. Instead, a chunk of granite that could easily have killed him "chewed me up and sent me home by chopper," wrote Mantle of the incident in 1998, when he would become the first climber to complete the SPS list solo.

A few hundred feet from the summit, Mantle—who'd already climbed thirteen peaks in the previous ten days—reached for a hold on an estimated 500-pound boulder, and it came down with him. "After tumbling about thirty feet," wrote one of Mantle's companions, Tina

Stough, in an article published in the *Sierra Echo*, the SPS newsletter, "he landed upright, sitting down hard on a jagged protruding rock, his right foot trapped between this rock and the main culprit as sand filled in from above."

Blood poured from a "huge gash" below Mantle's right knee, but the bleeding was stopped by one member of his team while another ran 5 miles to the backcountry ranger at McClure Meadow for help. Despite numerous lacerations, bruises, and possible broken bones, the hardy climber, whom Stough referred to in her article as "the new Norman Clyde," remained conscious, reciting T. S. Eliot's poem "The Love Song of J. Alfred Prufrock" and dialogue from *The Music Man*. "What a trooper," wrote Stough.

As luck would have it, ranger Em Scattaregia was at her cabin and initiated the SAR. Mantle's location "was reported as 300 feet from a suitable landing site, on somewhat loose talus, but not requiring ropes," wrote Scattaregia in her logbook. "Turns out the report was erroneous and the patient was on class 3 terrain in a narrow chute, 600 vertical feet above a landing site. Called in Yosemite helicopter which had short-haul capabilities. They couldn't get him after two tries due to strong and variable winds."

By 4:30 P.M., Randy was flown in to help Scattaregia and two park medics trained in technical rope rescues. Mantle's condition was stable, but the condition of an injured person can change quickly at altitude, so getting him down to a lower elevation for an immediate pickup at first light was paramount on everyone's mind. The operation would therefore have to be done in the dark, by headlamp.

Of main concern was the possibility of rocks falling on the rescuers as Mantle was lowered in stages down the chute. Five hours later, he was in a suitably safe location, a talus field near the base of the peak. It was 12:30 A.M. and Randy was exhausted. The day before he'd patrolled nearly 20 miles, then stayed awake that night tending to Ingraham and preparing for Hoffman's SAR.

Mantle bivouacked with two of his climbing companions, while Randy huddled with the medics and the remainder of Mantle's crew around a small fire, trying to keep warm as temperatures dropped into the low twenties. At 5 A.M. the group finished transporting Mantle to a landing site. By 9 A.M., Mantle was in a helicopter being flown to a hospital in Bishop.

Four park rangers were used for this particular SAR. In her article in the *Sierra Echo,* Stough noted that "Five helicopters had been used over the course of the rescue, and we did not have to pay a cent since we were in a National Park. Many thanks to the NPS for helping our pal Doug!"

IN THE LATE 1980S AND EARLY 1990S, a few dozen business cards circulated quietly among rangers during training. They read:

Dear Park Visitor:

You've just had your pudgy and worthless ass hauled out of deep doo-doo by a bunch of underpaid but darned dedicated public servants employed by the National Park Service. This mission was accomplished using outdated and rickety equipment made to work by a child-like faith in duct tape. But even duct tape costs money and we don't have much of either. Your sunglasses cost more than we make in a week. How about spreading some of that wealth around and contributing to our Search and Rescue Fund?

Thank you,
Your National Park Service

Ah, ranger training. Some called it charm school, others merely groaned. For backcountry rangers, this was the time for bonding, because once they were flown in to their duty stations, face-to-face socializing would be nearly impossible for the next three months, likely occurring only during search-and-rescue operations or on the rare occasion two rangers met on patrol.

Over the years, ranger training had escalated from none to a laundry list of requirements. Usually it kicked off with a welcome meeting in which the park superintendent or chief ranger provided an overview of the state of the parks and rallied the troops—a pep talk that, according to the rangers, almost always ended with "Budgets are tight, we're glad you're here, bear with us, maybe it will be better next year."

Then the frontcountry rangers went one way and the backcountry rangers the other—to attend a week of courses that taught them usable skills such as resource management, radio shop protocols, swift-water rescue, helicopter safety procedures, technical rescue, and emergency medical technician refreshers that covered, for instance, newly adopted CPR techniques, identifying high-altitude pulmonary edema and cerebral edema, administering oxygen, and finding a vein and starting an IV.

Frontcountry and backcountry cadres reconvened the following week at law enforcement training, which included such courses as Firearms Qualifications (aka target practice), Law Enforcement: Rangers' Roles and Responsibilities, and Physical Fitness. Vague titles, such as Gangs, had an obvious urban crossover theme, while Historic and Prehistoric Artifacts, Crime Scene Investigation, kept even the backcountry rangers awake. The related video, *Halting Thieves of Time, Protection of Archeological Resources*, seemed equally worthy of a cup of coffee before the lights dimmed. Other requirements—Defensive Tactics Training—and videos—*Stress Shooting, Mental Conditioning for Combat*—were the types of classes that made Randy nostalgic for the old days. He preferred courses like Verbal Judo, which taught rangers how to peacefully talk down aggressive individuals without the use of physical force. But times had changed, and so had ranger training.

Randy was one of the only rangers in Sequoia and Kings Canyon who remembered when there was no training—a time in the 1960s and 1970s when guns weren't a mandatory part of a ranger's equipment list. In 1965, "the job was to hike the trails, talk to people, send dogs out of the backcountry, write fire permits, clean campsites, put up drift fences, and when people needed help, call for it," wrote Randy.

"There was virtually no preseason training. If there was a position description I never saw it. Not even a first aid card was required." Very loosely stated, his and other rangers' duties in the early years were "to protect the people from the park and the park from the people."

In 1978 that code expanded to include "to protect the people from the people." That was the first year a ranger was required to have a law enforcement officer (LEO) commission in order to give citations and make arrests. More importantly, at least for Randy, was that an LEO commission was necessary for a "long season," which extended a summer ranger's job—usually ending in September—by a month. October was when he roamed the park boundaries on "hunting patrol," looking for poachers who had "wandered" into the parks while hunting deer. More time in the backcountry was always a good thing to Randy.

Weeks spent in the classroom took away from time in the backcountry, and Randy resented that. He wasn't alone when he shook his head and said, "What is this shit?" when, for example, he had to be certified as a "breath-test operator" after successfully completing a course in the "theory and operation of the Intoxilyzer 5000" taught by a forensic alcohol analyst from the California Highway Patrol. The backcountry rangers, who became known as the "backbenchers," grumbled audibly. Understandably so. This test was used almost exclusively on suspected drunk drivers, and there weren't even roads, much less automobiles, in the backcountry. What it came down to was that the rangers were required to attend 40 hours of pure law enforcement training—not SAR or EMS; it had to be law enforcement—even if it was completely useless for their jobs. If Randy was undertrained at the beginning of his career, by the 1990s he was thoroughly and completely overtrained.

Randy learned to cope with the inappropriate "required" courses by poking fun, irreverently, along with other backbenchers—including Walt Hoffman, who arrived late to a Gang Violence class and announced loudly to the half-asleep throngs, "Is this the meeting of the Gay Rangers for Christ?" The class broke out in laughter. In another class, while a graphic crime-scene image was being projected on the wall, one of the backbenchers chimed in sarcastically, "Well, everybody needs a hobby." A frontcountry law enforcement ranger on loan from the Los Angeles Police Department looked back from the front row and said, "You guys must be the old hands. They always sit in back." After class, the ranger instructing the course, a permanent named Eric Morey, walked up to the little clique of backcountry rangers, the palms of his hands almost touching each other, and said, "You guys were this close to redlining my pissed-off meter back there."

In many classes, the backcountry rangers simply wrote letters or tried to avoid the "nod-and-jerk boogie" that came with falling asleep while sitting up. One season they passed around *A River Runs Through It*, taking turns reading the book. "But Randy didn't take part in that," says George Durkee. "He would actually sit there and look like he was listening—probably in deep meditation, or, I suspect, wandering along a mountain stream." In fact, when training got to be too much—it sometimes lasted two and a half weeks—Randy would place his hands on his knees, in an impromptu lotus pose, and begin chanting, "Gentian, gentian, gentian," after the mountain flower, "a reminder," says Durkee, "of why we were there."

There were also classes wholly relevant to rangers' jobs in the backcountry. Randy was attentive during classes covering the law when it came to such things as a warrant in order to search a tent or probable cause.

In spite of all the ribbing and sarcasm, the backcountry rangers were good at what they did. Granted, the frontcountry law enforcement rangers wouldn't pick a stereotypical backcountry ranger as their first-choice backup in an armed confrontation with a drunk camper in a Winnebago, just as the backcountry rangers wouldn't choose fresh-from-the-city law enforcement frontcountry rangers to belay them down a cliff—or participate in a search-and-rescue operation, for that matter. "They could smell a joint from a mile away, but they couldn't find their way out of a dime bag if their life depended on it" was one of the more classic descriptions of a frontcountry ranger in the backcountry.

Some considered the backcountry rangers arrogant. They generally associated only with each other and made little effort to talk to the other rangers at training. Their behavior wasn't endearing. "My only defense to this," says Durkee, "was that after a certain number of years, these other rangers—permanents and new backcountry rangers too—would come and go. Many wouldn't last the season. It was hard to expend the effort to talk to them. At some point, though, it dawned on us that a vaguely familiar face kept coming back and might be just as excited about the park as us. Maybe even worth talking to."

It was because of this keep-to-the-group mentality that many of the backcountry rangers, and certainly Randy, earned reputations as being reclusive. "They weren't outwardly mean or anything," says Scott Williams, a ranger who experienced their vibe when he started out at Sequoia and Kings Canyon in the 1980s. "They just kept to themselves—and for some that created resentment, but for me, that created a mystique. Especially Randy, who was the classic mountain man. So one time I decided I was going to hike in and sort of invite myself to stay at his cabin at Charlotte Lake. I'll never forget the greeting I got. Randy was walking up from the lake, carrying two heavy buckets of water, and I introduced myself on the trail. Said hi, held my hand out to shake, and his response was 'A lot of work for Giardia water,' and kept right on walking. But to be honest, it didn't take a lot to earn their respect. All you really had to do was show that you appreciated the wilderness, jump in and pick up some trash, pack it out, and if you stuck around long enough, they eventually warmed up."

Eric Morey, whom Durkee says "came to respect us and even liked us, in spite of ourselves," began a tradition of inviting the entire backcountry crew to his house for dinner one night during training. "Randy, me, Terry Gustafson, Bob Kenan, Lorenzo Stowell, Dario Malengo, and Lo Lyness were there sitting at one table," recounts Durkee, "and the chief ranger, Debbie Bird, leaned over and told Morey, 'There's got to be more than a century of backcountry experience sitting there.' "

Actually, it was around 130 years of cumulative experience, with Randy at the head of the class.

After the summer of 1989, Alden Nash sat at his desk and reviewed Randy's twenty-first season. He had never received a legitimate bad mark on a performance appraisal form, and this season was no exception.

"As the senior member of our wilderness staff," wrote Nash, "Randy's perspective and work ethic is a model for others to follow. He works well in remote wilderness stations far from direct supervision and with difficult lines of communication. On his own time and expense he has kept both Law Enforcement and EMS certifications current over the years. His rapport with fellow employees, park visitors, and supervisors is excellent. He continues to maintain a focus on the National Park Service mission while living and working under third-world conditions. Randy's paperwork and reports indicate a thoughtful caring attitude towards the job and wilderness. It is an honor to have Randy on our wilderness staff."

CHAPTER NINE

GRANITE AND DESIRE

Rangers . . . are no different from other men, with
the same problems and burdens, the same urges and
conflicts, and the same vices and virtues. In other
words, being rangers does not keep them from being
just men.

—*Jack Moomaw,* Recollections of a Rocky Mountain
Ranger

Here [in wilderness] destruction for our recreational
pleasure is bad. . . . Stealing is bad, for that always
injures. Fornication is not when it affords pleasure for
all. But sex with a married person would be a bummer
if the individual's spouse is hurt by it. Which begins to
make the good-bad question complex—it nearly always is.

—*Randy Morgenson, McClure Meadow, 1973*

BY 1990 RANDY was comfortable with the unofficial but widely accepted opinion that he was the most fanatical environmentally conscious ranger at Sequoia and Kings Canyon National Parks—in the
entire National Park Service, many speculated. He'd adopted Edward
Abbey's term "syphilization" when describing civilization; when his

fellow rangers made comments like "When I get back to reality," he'd correct them and say, "Hey, this *is* reality."

For Randy, park headquarters at Ash Mountain was "Trash Mountain," and happiness was "Trash Mountain in your rearview mirror."

He was a stone's throw from 50, and with twenty-three seasons under his belt, he had "seen some shit." The "and there we were" stories were endless. He'd been bluff-charged by bears, rescued damsels in distress, returned missing Boy Scouts to their worried parents, lowered climbers off game-over cliffs, all the stuff of ranger lore—but those were the stories he wrote the least about in his station logbooks and personal diaries. A search-and-rescue operation might get two sentences, while the song of the hermit thrush would get two pages.

Randy wrote long about wilderness—so long, in fact, that park administrators taped notes on the covers of logbooks that read, in bold type, "Please avoid the James Michener Syndrome." To which Randy editorialized beneath, in his neat handwriting: "Got somethin' against literature?" He never really took Wallace Stegner's "almost-infallible rule of thumb" to heart, that "nature description by itself is . . . pretty inert and undramatic." Not so for Randy. Anybody who read his logbooks understood that protecting the people from the park and the people from the people was his job, but protecting the park from the people was his life's work and his passion. As Rick Sanger puts it, "It's not enough to say that Randy loved the Sierra. His soul had grown deep roots right into the sparkling granite of the place."

"We are children of the Earth, much more than this civilization wishes to admit in spite of our bulldozers and cement plants," Randy wrote in 1972 while stationed at McClure Meadow. "We can deny this only thru an ingenious self-delusion, and delusion is never honest or healthy. As we turn away from the natural Earth, we turn away from a vital part of ourselves. Our health declines.

"How have I understood these things? Not thru any sort of logical reasoning, but through stillness and quiet on an alpine lake. I've felt an honest, wholesome goodness within. I begin to realize that these native places are vital to my completeness as a man.

"I don't use place names in order to protect these innocent places.

"A white gull on a high lake, a dipper in a tumbling, noisy canyon stream—a bird at home with water. To fly and swim. What a grand existence that would be! Man is a poor creature. How clumsy we are in our own element. What land creature as highly developed as man struggles about over its surface as we? How many years for us to learn to walk erect? And now what do I hear most? Blisters and sore feet."

While at McClure Meadow in 1990, Randy wrote on a loose piece of paper: "I live in a valley at 9,700 feet in the High Sierra. I won't tell you where it is, for what I have to say about it may entice some of you to come, and there are enough already. Fortunately many of you prefer your screaming, blackened sulfur dioxide cities. Splendid! Let not I be the one to draw you out. The more of you who remain, the more lonely will be my mountains, which is just the way I prefer them. Nor would I tell those of you who are seeking this country where I live. Find it yourselves, and it will be all the sweeter."

MORE THAN ANYTHING, Randy wanted to make a difference. That, plus his minimal salary, kept him semi-satisfied for a quarter of a century. Living in the high country was the real reward. Still, after all those years of service, he would have appreciated more than the two monetary rewards he had received, one for valor in a rescue on Mount Darwin, when he got clocked on the head by a falling rock (the chief ranger at the time at first denied the supervisor's request for that reward, reportedly stating that he was "just doing his job"). Durkee even sent a letter to the subdistrict ranger once, singing Randy's praises. He wrote, the "NPS, and especially Sequoia and Kings Canyon, does an abysmal job of recognition for seasonals—at Ash Mt. if you [were a permanent] and came to work with your shoes tied an award was in order."

Sounds like sour grapes—but many higher-level administrators agree with Durkee's assessment, saying such things as "Seasonals are treated like second-class citizens; they do most of the work and get the least recognition" or the catchall "Seasonals are treated like shit." One permanent employee who had worked his way to a high-level position in an NPS regional office offers this: "The National Park Service doesn't promote people, they promote egos. An ego implements, for example, a menial job like cleaning up a backcountry region of trash; the ego assigns the work to seasonal rangers, and then he puts on his résumé that he was responsible for 'clearing more than 2,000 pounds of garbage from the backcountry.' That same ego will sputter and hiss when asked to sign an overtime sheet for one of those rangers sweating in the field who got called to assist a backpacker on his sixth day." More diplomatically, Subdistrict Ranger Cindy Purcell says Randy's dedication "wasn't recognized by the Park Service. Not like it should have been after all that time. The system just doesn't account for seasonals like Randy."

Administrators, most of whom haven't received much recognition themselves on the way up the NPS ladder, will often tell you seasonal rangers are the backbone of the national parks. Yet there is no official length-of-service award or commendation for seasonal rangers. Permanent rangers—who aren't treated like royalty either—can at least look forward to ten-, twenty-, thirty-year pins, the kinds of tokens of appreciation that Randy desired.

"Where's the recognition for time in federal service for seasonal employees?" Randy voiced to the administrators in his end-of-season report in 1993. "Many such awards were published in the *Gigantea* [SEKI's employee newsletter] this summer for permanent employees. I have ten plus years total federal time, 26 seasons with this park. . . . Jack Davis as Superintendent initiated a seasonal pin as award recognition but he's gone and perhaps that as well (or have I just missed it?). But how about top level agency recognition to seasonals for years in service (either seasons or total years or both—in a quarter century to accumulate ten years of service, and dedicating oneself to NPS as a seasonal for a quarter century or more is quite an accomplishment)? There are a number of us who qualify at least for the standard agency ten year awards."

He needed to know that his presence and job were worthwhile, but in the 1990s he reflected back and could think of only one time that something he'd suggested had actually been acted upon—in 1982, when he'd argued not to upgrade the Shepherd Pass Trail for easier stock access into the park, citing the negative impacts on the native bighorn sheep and meadows that were contrary to the backcountry management plan for minimal use. Another time, he took matters into his own hands and bribed a government mapmaker—who was in the mountains to verify trails—with pancakes to delete an older trail off the next U.S. Geological Survey printing of a quad map in the LeConte Canyon region.

It was frustrating, because Randy and his fellow backcountry rangers lived in the mountains, yet their voices about the mountains, he felt, were rarely heard. Programs for the backcountry were rarely "in the budget," so the rangers—*all* the rangers—followed a strict set of rules regarding overtime pay and helicopter use for unofficial park business. Those same rules didn't apply to park administrators, or their friends, or local politicians whom they allowed to bend the rules, to the detriment of the wilds.

On June 24, 1976, when Randy was stationed at Tyndall Creek, a politician had asked to bring a group of what Randy estimated were thirty-five to forty people into the Rock Creek area. If that estimate was correct, Randy calculated, that many people would require sixty to seventy head of stock, far beyond the limit of twenty per group. To Randy's chagrin, the park superintendent had allowed it. "Once again," wrote Randy in his logbook, "as so often in this Great Nation, politics prevails over all other values. A county supervisor is far more important than a mountain meadow, or consistency within the law (would the Superintendent have granted permission to Joe Schmaltz? No . . .), or the backcountry experience in this park of those numerous small groups of backpackers certain to be camped up and down Rock Creek in late June. A fitting way to celebrate the much-heralded bicentennial."

Bashing the local politicians wasn't enough; why not direct a few choice words at the parks' superintendent? "I hear on the morning news that the Superintendent and his staff, on the third day of their one week backcountry trip, are requesting the helicopter bring them

Bisquick, cooking oil, and corn meal for their fish, to be delivered to their camp at upper Funston Meadow on Wednesday," wrote Randy on August 14, 1978, while stationed at Little Five Lakes. "Pretty nervy in a park where there is so much upper level flack about backcountry helicopter use. How'll he explain this landing in the Kern Canyon to the other campers at that meadow? Maybe we should start a resupply service for backcountry visitors, for to be just those who are paying should get a piece of that machine also. And rangers in the backcountry for three months can't get a string bean via the helicopter. Thus, a few words about helicopters for The Committee, that amorphous, anonymous, amoeboid body which sits in ultimate judgment in all communist states."

Randy never held anything back in his "rarely read by anybody important" logbooks. However, he toned down the end-of-season reports because *someone* might actually read them. In 1989, Randy typed a sixteen-page EOS report after having spent the summer at McClure Meadow. It was probably the most detailed EOS report to ever land on a subdistrict ranger's desk—"with a thud," remembers Alden Nash, who collected such reports and pushed them uphill "with a pointed stick" to the chief ranger, who theoretically sifted through and forwarded on suggestions to the superintendent. If anything seemed worthy of national policy change, the superintendent would deliver the recommendation via his own annual report to Washington, D.C.

"If anything a backcountry ranger suggested made it to Washington," says Nash, "it would be a miracle."

Randy, however, had an ace in the hole in 1989. Chief Ranger Doug Morris had visited him at McClure—on foot, no less—and was in agreement with some of Randy's ideas. Randy thought Morris was a stand-up character, and he sprinkled his name liberally throughout the EOS report—a culminating crescendo of two decades' worth of built-up frustration and anger.

Anybody who had known Randy for any amount of time could have guessed the overriding theme of this report before they even opened it.

Meadows.

Meadows, as Randy clarified in his report, "are not pastures. Their grasses and sedges are not feed. Managing for sustained yield is not our business. Managing for natural processes is our stated business, and as Doug [Morris] says, a grazed meadow is an unnatural situation."

Randy was referring to the use of stock in the backcountry, the impact of which had been a heated philosophical debate between packers and environmentalists for years. Randy devoted four pages to such "grazing" issues.

As he had virtually every year since he'd first been stationed there, Randy requested that McClure Meadow be closed to grazing: "It is a very special place and numerous comments by hikers support this. I would like to see our management policies support this. With several meadows and abundant woodland forage in Evolution Valley we can surely preserve one ungrazed meadow. Over 95 percent of the visitors are hikers and they have little opportunity to see an ungrazed meadow. The arguments for this protection may not be based on grazing-damage data, but the argument . . . is certainly emotional. . . . In any case, perhaps the quest for data to support our actions gets overemphasized. After all, our emotions distinguish us. Art and poetry and music are from and to the human heart, as is, for many, our relationship with the land. There has been a good deal of philosophical and emotional response to landscapes embedded in the conservation movement from the beginning.

"All the meadows in Evolution Valley were grazed this summer, and they all looked it. Yet Franklin Meadow apparently was not, and in October it was a place of knee-high grasses, ripe and open panicles drifting on the moving air, luminous-bronze in the backlight. It was a very different place and a very different emotional experience of a mountain meadow, and entirely consistent with what one might rightly expect of national park backcountry. It was a garden. I sometimes wonder whether range management concepts are any more applicable to our business than timber management concepts. The difference between a grazed meadow and a logged forest may only be one of scale."

Randy then inserted a hint of diplomacy: "For the stock user [clos-
ing McClure Meadow] means asking him to change some habits, to
think more of grazing woodland forage rather than prime meadows,
and even think of carrying supplemental feed. But it does NOT mean
a first step toward excluding stock from the backcountry. . . . We have
made a commitment to stock users to allow them to visit these mountains. But that need not mean they can graze any meadow they want,
as much as they want, until we can prove with facts and data they are
causing long-term ecologic change.

"We can protect them on our own long-term tradition of protecting
particularly beautiful places.

"Doug [Morris] is right about something else. Stock users have
been disproportionately vocal, and hence influential in our planning
process. There is no doubt in my mind that were everyone who gets
a wilderness permit allowed to vote, yea or nay, on the question of
stock in the mountains, stock would be gone. Stock users are a small
minority. Perhaps in an alleged democracy this is the way it should be
done."

Randy then posed a direct challenge to Morris and the superintendent: "I hope we have a chief ranger and superintendent willing to
stand for greater protection for more mountain meadows as wholly
consistent with NPS mission, to resist pressures for use, and to resist
the argument that we need prove and document this long-term change
thing before we can regulate use. That is not the only standard available to us.

"Should any of this planning move forward it would seem there is a
role for the backcountry rangers. I can't imagine anyone in the administration who knows the backcountry as these people do, yet they are
out of the planning and decision making loop."

Randy suggested that the administrators "take a more professional
approach to the backcountry ranger job. We are assigned a daily 8-hour
shift, yet the reality is day and night availability 7 days a week. There
is an enormous amount of unpaid overtime all backcountry rangers
work during daily patrols. And living in a ranger station is being con-

tinuously on-call and available to the public, which is how NPS wants it. People arrive at any hour. They have opened the door and walked into the cabin while I was bathing. Meals are regularly interrupted. Shelter is expected during storms. It goes with the territory. The reality of the job is that it is unworkable to set an 8-hour shift and expect no work outside those hours. Things can arise at any hour."

Randy covered a few scenarios, including "standby pay" for those inevitable duties during off-hours. But "I have been told this could be cost-prohibitive. Unless I misunderstood what I hear of the parks' budget, the safety budget has increased about 20,000 dollars, about 10,000 dollars will be spent to go to a parade in Bishop, and Sierra District created a new GS-9 position. Money appears to be available.

"If money for standby pay cannot be found, I suggest that for 1990 there be a decision made about the backcountry ranger shift and no work be asked or expected outside those hours without pay. No radio calls, no helicopters, no hiking anywhere, no employer visits, no visitor services. And I suggest there be established a system of recording and paying overtime for all those extra hours when a ranger unavoidably provides visitor services outside the paid shift. To continue to fail to do these things is certainly unprofessional, and probably illegal."

The end of the report clearly asserted Randy's yearly suspicion: "I would be interested in seeing the comments of those who read this. I have never known what anyone thinks about these annual reports." As an endnote he added, "Report composed to R. Strauss' *Death and Transfiguration*, October, 1989." The significance of the endnote was probably lost on anybody reading the report. German composer Richard Strauss had written the symphonic tone poem *Death and Transfiguration* in 1889, one hundred years before. In it he conveyed the inner thoughts of a young man struggling to accept his own death. The young man was an idealist who was struck down by a terminal illness, thus losing his ambition and will to fight for the causes he believed in.

Randy often listened to such classical music while writing reports. The somberness of *Death and Transfiguration* no doubt mirrored Randy's mood at the time. After more than twenty years of lobbying

without success to close certain meadows, including McClure, he was
losing his will to fight.

IN MAY 1992, Esther was 83 and becoming increasingly frail, so after
twelve years of living in Susanville, Randy and Judi sold their home and
moved to Sedona. While Randy was in the backcountry, Judi shopped
for and took care of her mother-in-law. Shortly after Randy returned
from the mountains, Esther fell and broke her femur. A bone scan revealed advanced-stage cancer.

Randy moved in with his mother immediately and "never came
home," says Judi. He attended church with Esther—something he
had quit doing after the Peace Corps, except during holidays with
family. Once she became too ill to leave the house—the cancer had
spread throughout her body, including her brain—he spent months at
her bedside. As her condition worsened, Esther withdrew and spoke
only occasionally. When Randy told her how much he loved her, she
wouldn't reply. He tried to get her to talk about her feelings and her
thoughts, and she said, "You want me to talk to you? What should I
talk about?" Randy's response was "I just want to know what I can
do for you, Mom." She replied, "You can't do anything, so I guess you
shouldn't try."

Randy sometimes read books aloud to Esther, who would listen for
a while and then become agitated and shake her head for him to stop.
Randy felt more and more helpless. "Esther did not want to die," says
Judi, "and she was angry because of it." Randy received the brunt of
her displeasure.

Larry came to help a few times, but he always seemed in a rush to
get back to New Mexico. And besides, he had medical problems of
his own—seizures that Randy suspected were due to alcohol. Randy
resented his older brother for not being there, yet when he was there,
Randy couldn't wait for him to leave. They'd long been estranged. For a
short time after their father's death, they'd spoken of taking a trip into
the high country in his honor, but that had never transpired.

Though Randy had often complimented his brother on being both smarter and more outgoing than he was, it worried him that Larry "liked to push the limits and didn't listen to caution," says Judi. "Randy became embittered over time because Larry always came to Esther to bail him out of financial situations." Gone were those enchanted times in Yosemite: ice skating at Curry Village; Randy beaming at his brother as Larry led the conversations at the dinner table; fishing in the Merced; and those magical wildflower walks with their father. Randy's first wish on his twenty-first birthday had been to go to the Ahwahnee Hotel and buy a beer—not for himself, but for his big brother. But here in the Arizona desert, those days were a distant memory.

When Esther passed away on April 22, 1993, Randy had taken care of her for almost seven months—and almost exclusively, except at the very end, when Judi had suggested they get hospice help. Randy was exhausted, emotionally drained. "Empty," says Judi. "He needed to get back into the mountains."

But first there were details. Larry had a strong attachment to the house and wanted to move in, even though he didn't have any means of supporting himself. Esther had already distributed her most valuable items, such as the Navajo artifacts and jewelry she and Dana had bought together—rare, museum-quality pieces. Beyond that, there were standard household wares, and the Morgenson family library. They all decided on an estate sale service that swooped into the house the following week, tacking price tags on everything in sight. Judi remembers Larry wandering around saying "Oh dear" and then breaking down in tears while standing amid the chaos. Randy was more stoic but no less emotional, with a "lump in his throat" through the entire process.

The family photos, writings, and library had been collected in one room that was out of bounds for the estate sale. Randy and Larry, harmonious for the first time in years, sat down together and divided the stacks of glorious books. Everything else was sold. Larry, with money in his pocket, bought out Randy's share of the empty house.

Esther had expressed her life philosophy to Randy more than once. "There are times occasionally in life when great changes occur," she wrote to him when he left for India, "and then nothing is ever the same again. Things never stay just the same, any time. Change seems to be one of the few certainties in life. Just as well. But while we look forward eagerly to what is to come, we can thoughtfully appreciate the good that has been and what we have at the moment."

With a heavy heart, Randy left for training.

RANDY ARRIVED at training the summer of 1993 understandably numb—fallout from the previous months at his mother's death-bed—but excited to be among friends and in the shadow of his mountains.

During those six months, he had undergone a change. Perhaps Esther's death had served as a reminder that life does end, and if there was anything he wanted to do, he had better get on with it. But it was complicated. All his life he had struggled with what society expected of him. He was a dreamer and had a hard time manifesting those dreams beyond the High Sierra. The rebel in him thumbed his nose at convention, but he also had been brought up with deeply instilled family values. He began to wonder if he wanted to remain married. He wondered if he and Judi were compatible after all.

Randy had always wanted Judi to spend more time in the mountains with him. As the years progressed, it seemed as if she was finding ways to spend less. She'd teach summer school or cut her trips down to only a week or two instead of the four she'd spent at the beginning of their marriage. Usually, however, she was dealing with some crisis in her own family. Over time Randy came to resent his in-laws because the crises always seemed to hit during the weeks—the only weeks—Judi and he could be together in the wilderness.

Despite the romantic façade, the backcountry wasn't always a honeymoon for the two of them. "We had some rough times when I came to see him . . . some of our worst fights, sometimes because he pushed me too hard or because of the cramped quarters of the cabin or tent, and sometimes just because it took a while to readjust to each other after being apart," says Judi. "And I wasn't always Florence Nightingale." Randy was hard on all his guests—he expected more from them than he did from the backpackers who passed through his territory. He could be particularly tough on Judi. Once, the sole on one of her worn hiking boots nearly fell off during a patrol, and Randy berated her for the oversight. He couldn't believe she hadn't thoroughly checked her footwear in advance. In the cabin, if she hung the spatula on the wrong hook, he'd make a show of the error. "The backcountry was his domain," Judi says, "and he wanted it a certain way. He could be a real jerk."

There were, more often than not, magical moments: "We'd have moonlight strolls where the granite was lit up like daytime, and at dusk, Randy knew the perfect spot to watch the peaks turn red, orange, and pink. He always knew the perfect spot to sit, and he'd know the exact spot on a mountain that would get the last light." Randy hadn't seen fireworks on the Fourth of July for twenty-eight years, and he often said he didn't miss a thing, preferring the Sierra shows, calling them "fire in the sky." Randy would say it was Mother Nature painting the mountains with light. He was at his most romantic in the mountains—he could describe a dead foxtail pine snag as though it was a rose garden in Shangri-la.

Rangering was like pioneering. "Some of the stations had wood-burning ovens," says Judi. "Over trial and error, I finally stopped burning the bread." But Randy didn't want Judi to stay in the cabin and cook for him, and neither did Judi. They cooked together, patrolled together, and were happier when they were out in the wilds.

Judi had joined Randy in the backcountry almost every summer since they'd met. She'd joined him on trans-Sierra ski tours with the boys, where she'd proven herself an adventurous and capable mountaineer, but that summer, after his mother's death, Randy began to wonder if she would have done these things without him. Then he would loop back to her devotion, their mutual love of art, her patience with his job and his schedule, and her support—she both supported and encouraged him to do these trips in the wilds. Never did she tell him not to go. Did it really matter if the call of the wild didn't course through her veins as it did through his? The fact was, she usually enjoyed herself in the mountains. Sometimes she'd get scared, and whine, and complain—but she'd do it anyway. That was love.

Only Randy could say if any of these thoughts crossed his mind when he strayed from his marital vows and ended up in the same sleeping bag as Lo Lyness.

Lyness was one of the core crew of backcountry rangers and a teacher of outdoor education during the off-season. She had fallen "deeply in love" with the Sierra while working at a Sierra High Camp in 1976. Her wilderness career began with the U.S. Forest Service, but she left what she termed the "U.S. Forest Circus" shortly thereafter and climbed the NPS seasonal ladder, working in campgrounds, doing road patrol, and reaching her backcountry ranger goal in 1981. A Stanford graduate, Lyness could banter decisively on any number of topics, especially those concerning the environment. In the past Randy had written, "All summer long with everyone I meet it's impossible, with a rare exception, to relate as myself. They all relate to me, and force me to relate to them, as The Ranger. That's where the loneliness is. If I'm ever lonely it's for a friend, with whom I can speak as me."

Lyness shared Randy's passion for the wilderness—this was her twelfth year as a backcountry ranger—and a close friend of Randy's says that she "represented things that Judi did not." But if she envisioned any long-term plans, Lyness might have been setting herself up for disappointment. Randy's follow-up line in this journal entry was "In fact I'm least lonely when I'm all alone."

Their relationship, which began during a late-season EMT training course at Ash Mountain in 1993, became complicated because they hadn't been discreet with George Durkee, who felt that he'd been placed in an awkward position. For almost twenty years he had been a friend to both Judi and Randy. Durkee asked Randy point-blank, "What about Judi?" Randy launched into a long rationalization about not wanting to be constrained by "Western morality." To which Durkee responded—knowing that Buddhists are taught to refrain from harming themselves or others through sexuality—"So much for Buddha."

In academic debates Durkee and Randy had always been on an even playing field. In this case, Randy was left speechless.

After some silence, Durkee told Randy that he needed to make some decisions, because he wasn't going to lie to Judi, nor was he going to be burdened with having to tell her the truth. If the affair ended right then and there, Durkee decided, it wasn't his place to say a thing. A couple of weeks later, Durkee received a letter from Lyness.

"She wrote, that's not who I am," says Durkee. "She said she was ashamed and embarrassed, and it wouldn't happen again." Durkee wrote back and said, essentially, "Okay, shit happens. We have a pretty weird job, just a weak moment," and so on. Randy had conveyed to him essentially the same sentiment as Lyness had.

"Game over," thought Durkee.

MANY OF THE BACKCOUNTRY RANGERS, including Lo Lyness, had, over the years, spoken about one of them writing a book on all the "crazy" backcountry relationships. Somebody dubbed it *Granite and Desire*. They joked about it becoming a television soap opera, maybe even a major motion picture.

Of course, in the real-life *Granite and Desire* these weren't actors, they were real people, and real people are bound to get hurt. Lyness and Randy's relationship didn't end in 1993. In the fall of 1994, the burden of knowing about it was still planted firmly on the shoulders of Durkee and his wife, Paige Meier. A few astute backcountry rangers also had picked up on a "vibe" during training.

Durkee had already made it clear that he would not keep the secret if pressed by Judi, but Randy urged him directly, "Please don't tell Alden." Alden Nash had just retired in the spring of 1994, and he had become more than a well-liked supervisor to the backcountry ranger clan. For Randy especially, Nash was a very close friend, one of his confidants. In return, Nash had opened up his heart and his home to Randy on numerous occasions, going above and beyond his supervisory duties. Randy's direct request to keep his affair a secret from Nash spoke volumes. He wasn't proud of what was going on and, by all accounts, neither was Lyness. "The problem was," says Durkee, "they were falling in love."

On October 17, Judi called Gail Ritchie (now Gail Ritchie Bobeda) to wish her happy birthday, a mutual tradition the two had shared since the early 1970s after they'd traveled to Europe and worked in Yosemite together. Judi had always credited Gail—who now lived hundreds of miles away near Santa Cruz, California—as being the person who had introduced her to her soul mate, Randy. If Gail hadn't gotten her the job at The Art Place in Yosemite, they would never have met.

Judi tried to maintain a chipper attitude on her friend's birthday call, but in the course of the conversation, she started to cry. She told Gail that her back was out of whack and that she had been worried about Randy's behavior of late. To top it off, she hadn't seen him in months, and instead of coming right home when the season ended on October 5, he had gone to Yosemite and climbed Mount Dana, where he and his brother had scattered their father's ashes back in 1980. Now they had returned to the summit of that sacred family mountain to offer their mother's ashes to the wind. Judi had planned to join them far in advance, but at season's end, she told Randy over the phone that she was in pain because of back problems and was having a hard time getting around, and to come home as soon as possible. Here it was almost two weeks later, and he still wasn't home.

That evening, Gail went with her family to a club in Santa Cruz to listen to some live music and, to her surprise, bumped into Randy, whom Judi believed to be somewhere in the eastern Sierra. Randy "stuttered" when Gail asked him what he was doing in Santa Cruz, then said he was with "Park Service people." He asked Gail not to tell Judi that she'd seen him. Gail didn't agree to anything. It was obvious to her as she watched Randy return to his seat that he was with a blonde woman in the group—Lo Lyness.

Gail spent an agonizing night trying to decide whether to tell Judi or not. She believed in fate, and felt it was more than just a coincidence—and downright bizarre—that she had run into Randy hundreds of miles from Sequoia and Kings Canyon on the same day she had spoken to Judi.

She finally fell asleep after she made up her mind. She had to tell Judi.

The truth, delivered by Gail, broadsided Judi. She hung up the phone and immediately dialed George Durkee, who both confirmed the affair and apologized as Judi repeated tearfully, "I've lost my best friend. I've lost my best friend." But she wasn't entirely surprised. She'd been putting together clues for some time—mainly letters written by Lyness, aka "just a friend," that had begun the winter after Esther's death. Around the same time Judi noted that Randy's creativity had taken a serious dive. Once upon a time, the Museum of Art in Reno, Nevada, had requested twenty of his images for an exhibit and he had spent three months meticulously choosing the photographic tribute to the High Sierra. Now he had, for the most part, stopped taking photos. The dream darkroom he'd planned to build in the basement of their home was perpetually not under construction. Whereas he had once spent the winter processing the summer's photographs, researching environmental issues, and writing pertinent representatives in Congress, he instead moped around the house, ran endless miles, went on long hikes alone, and seemed happy only when he was gearing up to leave for the snow-surveying job he'd scored outside Bishop.

A woman knows, at least upon reflection.

The timing couldn't have been worse. Judi's mother had just been diagnosed with lung cancer. This after helping Randy take care of Esther as she battled cancer and, shortly before that, the death of one of Judi's brothers in 1992. She wondered what she had done to deserve this.

When Randy showed up with his tail between his legs, Judi was determined to be strong. She informed him of her mother's cancer and told him she was going to her mother's house and that it would be nice if he was gone when she got back. He maintained that he wasn't sure if he wanted to leave. Judi, who was deeply hurt and at an emotional disadvantage, eventually forgave him, but she didn't forget.

Randy then confided in the only person he could share his dilemma with—his photography partner from Susanville, Stuart Scofield. Scofield had once experienced similar turmoil in his own life and was determined not to turn his back on Randy or take any moral stance.

After Judi returned home from a long visit with her mother, Randy wrote Scofield:

> *Thanks for your notes. It means a lot to be remembered. We're healing ourselves and looking toward our future. Of course, everything will eventually be fine. Glad to hear your workshop load is increasing. Maybe that'll all come together for you now. I'm anxious to see some prints, in need of inspiration.*
>
> *—Cheers, Randy & Judi*

That tone didn't last. As Judi juggled her ceramics teaching schedule at a Sedona gallery with visits to her mother as she underwent chemotherapy, Randy worked on his head. The books he read that winter were found not in the nature section of the bookstore but in the self-help aisle. He was trying to understand why he had done what he'd done, why he was no longer content with his life, and why life itself had lost its luster. He was depressed and searching for answers. All he really looked forward to was going back to the mountains. He and Judi were cordial—she needed time to trust him again—but certainly not back to normal.

THAT SPRING, Randy continued to see Lyness while on snow surveys in the Sierra while maintaining a strained marriage with Judi. One day he would tell Judi that he thought his life was "slipping away," that he wanted to go to wild places she wasn't interested in—the classic midlife crisis. The next day, he would talk about quitting the Park Service and nurturing their life and marriage in the Southwest. They went to a marriage counselor, who spoke to both Judi and Randy separately, after which the counselor told Judi to "get out of the relationship" because they weren't compatible. Judi told Randy, who fumed, saying the counselor didn't know what she was talking about, that they were "extremely compatible." These vacillations continued, so when it came time for Randy to return to the mountains in June 1995, Judi wasn't ready to pull the plug on the relationship. Right then, she had a very sick mother to take care of. Randy told her he wouldn't go back if that was what she wanted, but then he'd say, "a season would really help me sort things through." Judi knew that Randy was going to be near Lyness at ranger training—and she just had to trust him. She kept telling herself that, but even though Randy had promised the affair was over, she was not reassured.

Before heading into the backcountry on June 4, 1995, Randy wrote Scofield again:

Hi Stuart,

I appreciated lots your birthday card. Thanks for thinking of me. You get into my thoughts also. Wish we could do a hike together. I've done no photography this winter—distressingly. But a recent couple of boxes of slides came with some happy surprises, so maybe it's still there. I can feel now what you went through during your turbulent years. Losing things, not doing your work. . . . Something like three years now without making a print. And nothing in sight. I need to learn to be content with incremental progress.

At training, Randy and Lyness were circumspect, even avoiding sitting next to each other during classes. But except for newbies like Rick Sanger, everybody knew what had transpired between the two rangers, and Randy took the time to walk around and apologize to anybody whom he felt had been made uncomfortable by their actions. He also asserted that he had patched things up with Judi. George Durkee sensed otherwise and confronted Randy. They had a small blowup, and Randy admitted that he'd considered suicide as a way out of the mess.

"Really?" asked Durkee.

Randy replied, "Not seriously, but I've been having those kinds of thoughts."

"You're certain?" said Durkee. "Not seriously?"

Randy said, "Yeah. I'm fine. I'd tell you if I wasn't."

As training continued, Randy had a lot on his mind and no time to chat with or even acknowledge new faces like Sanger, who was excited about his first season as a backcountry ranger. Randy and many of the backcountry rangers made Sanger feel unwelcome, even though Sanger was actually a "vaguely familiar face." He'd come to training two years in a row, hoping to get hired on. He'd gotten a law enforcement commission on his own, and his EMT certifications were up to date. All he needed was a nod from Randy Coffman, the district ranger, that the budget would allow his indoctrination into this elite crew, which seemed to be a completely unpleasant group of assholes who couldn't be bothered to give him the time of day.

Training ended, and Sanger was assigned a station. "It was one of the biggest thrills of my life," says Sanger, who was taken under the wing of an "extremely cool" backcountry ranger without an attitude, Rob Hayden. Sanger took the snubbing he had received as a challenge to crack the ranks of the other backcountry rangers. His personality was such that he found it very difficult not to like people—even assholes got the benefit of the doubt.

Randy was flown into LeConte Canyon on June 25, 1995. The mountains were still predominantly white with snow from the heavy winter storms. When Lyness returned to her station at Bench Lake, both the tent platform and tent were under 5 feet of snow. District Ranger Randy Coffman was flown in to inspect the situation. Unaware of Lyness's affair with Randy, he pronounced, "This will never work," and decided on the spot to send her to LeConte Canyon until the snow melted.

Unlike during the previous two summers, Randy seemed immediately taken by nature. "Aspens are just unfolding tiny leaves," he wrote in his logbook his second day in the backcountry. "Willow buds are swelling. Robins sing at dawn. Warblers in mating dance. Rangers in unpacking and cleaning dance. . . . River fluctuates about a foot from morn to eve. Snow must be recently gone for in wet places the warm earth just sends up fresh green shoots . . . the familiar comforting warmth of LeConte Canyon.

"Scrubbing, washing, cleaning day. Making space for supplies. 114 [Lyness] did a lot of scrubbing.

"Trying to make the psychic adjustment to being here."

For two weeks Randy and Lyness were, for the most part, alone in the snowbound LeConte Canyon. Seventeen miles north on the John Muir Trail, Durkee was at McClure Meadow. When he heard about the bunking arrangement at LeConte, he thought to himself, "Here we go again."

ON AUGUST 8, 1995, Randy patrolled to Bishop Pass and discovered that a bunch of cowboys from different packer outfits had teamed up, shoveling snow to allow stock access into the high country. There was a time when packers would sprinkle rubber shavings grated from auto tires across the snowy drifts along the trail, especially in the higher passes. The sun heated the shavings, which tripled the speed of the snowmelt. The NPS stopped the process shortly after Randy was hired in 1965, but thirty years later the occasional black shaving could still be found on Bishop and other passes.

Randy talked for a spell with one cowboy working a shovel near the upper third of the steep switchbacks—some of which have sheer drops to granite. The pass is a trail-construction marvel, a short, steep, dizzying climb. Those with a fear of heights usually hug the mountain side of the 2-foot-wide trail that is literally chiseled out of the granite. A stumble could send either animal or man to his grave—and has. Many a pack animal had fallen from these and other precipices. When this occurred, rangers, including Randy, assisted in disposing of the dead animal. This usually involved blowing up the carcass with dynamite, a grisly task that vaporized the remains and helped prevent the spread of disease to local wildlife.

Farther down the switchbacks, while chatting with two packers Randy knew from years past, a packer whom he did not recognize came marching toward them "about as fast as he could walk," wrote Randy in his logbook. So he "stepped back from the edge." The noticeably agitated packer "walked right up to my chest and snarled at me," wrote Randy. " 'Don't stare at me while I work. It pisses me off.'

We looked at each other. For a moment I thought he might take some further action, but he backed up, turned and slowly walked down the switchbacks. Seemed resolved for now so I said nothing."

Randy asked the other packers the man's name, but they knew him only as "Tom." That night Randy queried Laurie Church, who was near LeConte at a trail-crew camp. She said she had met the guy in previous years and he'd never seemed very friendly.

The next day Randy hiked out of the canyon again. At the rim where Dusy Basin begins, he encountered three people, two men and a woman, off the trail eating lunch. Without a word, one man handed Randy a wilderness permit. Randy exchanged some jokes with the other two backpackers and then, since the permit had been issued by the Rainbow Pack Station, he inquired where the packer was and where he was taking their gear.

At this, the man with the permit "explodes at me," wrote Randy. "He's angry because I ask questions ('Why don't you just ask the packer!') and he wants me to leave. ('I don't like you. I don't like rangers. I don't want you near me. I can't believe you don't leave when I've told you to. Etc.' shouted angrily.) Is he suddenly going to jump up and attack me? I try to ask what is the real problem (more, 'I don't like you.'), try to talk to him in the hopes we can calm this down. He gets angrier.

"Presently he began stuffing things in his pack, and said, 'I'm going to have to leave since you won't. You've ruined my lunch.'"

Alone with the other two hikers, Randy asked if he could talk to them for a few minutes. They invited him to sit. The man's friend "gently and politely tried to explain this guy sometimes has trouble with officious rangers," wrote Randy. "I don't doubt, with that kind of pugnacious attitude. So he just hates all rangers, and explodes at them. Then I learn this is the guy we rescued off the Hermit in 1988! A bunch of rangers, helicopters, a day and a night and into the next day . . . and all he can think is that rangers cause him grief! Ingrate! Whew!

"They are in now to climb on Devil's Crags. The day in 1988 I helped get this guy off the Hermit I spent the morning helping get a body off Devil's Crags.

"So . . . we aren't going to change anyone. People only change them-selves. There's no magic phrase to cause Satori. Only we can contain these things, prevent its getting worse, control it gently; and stand my ground—not be pushed, manipulated, or threatened successfully. Don't have to take it, but don't strike back. Firm, polite, resistance. Stand my ground."

Randy continued on his patrol away from his cabin, toward Bishop Pass, and eventually met up with Ed Bailey, the cowboy working for Rainbow Pack Outfitters who was leading three mules carrying the climbers' gear. They spoke for a while, and Randy commented in his logbook that Bailey was "pleasant." The gear was being dropped off at Grouse Meadow. Then, the packer told Randy, he planned to camp at Ladder Lake, using pellets he'd brought along as feed.

Randy was impressed.

The packers operating in the High Sierra had been lobbying to increase the stock limit from twenty head to twenty-five per group. Bailey admitted to Randy, "I'd rather stay with twenty; my pay doesn't go up for handling more."

"He could work for us with an attitude like that," wrote Randy that night.

Randy relayed the tense moments he'd experienced with "Packer Tom" three days earlier. Bailey supplied the packer's last name. Then Randy described the altercation he'd had just an hour before. As they parted ways, Bailey turned in his saddle and said what Randy described as an "encouraging word":

"I used to work at Harrah's. Sometimes you have to be 7 feet tall!"

Two days later, on August 13, Randy hiked to Grouse Meadow to check up on the three Devil's Crags climbers. He spotted a blue tent below the meadow and went to investigate. The tent was "torn up, fly ripped off, poles bent or broken, clothes bags inside torn open"—destroyed by a bear.

Randy wrote, "There were torn nylon bags on the ground and food residue and packages, wrappers, punctured cans, jars, fresh fruit, 2 wine bottles scattered around camp and into the grass and the wind starting to blow some of it away. This stuff isn't carried into the mountains by hikers; there were stock tracks into camp and there has been no stock over Bishop Pass and into the canyon except that packing for these climbers."

Via radio, Randy contacted a frontcountry ranger, who contacted Rainbow Pack Station and got the address for Doug Mantle, the signer for the backcountry permit. After weighting down the garbage most apt to blow away, Randy hiked back to his station. "Should have carried away the food mess," he wrote, "but it was much more than I could deal with in a day pack." That night, Randy was informed that Packer Tom had been fired for his behavior on Bishop Pass.

By season's end, Randy had met up with Lyness a handful of times, including what Durkee said was their "last patrol together" in Lake Basin. Combined with the two weeks at the beginning of the season, she'd spent more time patrolling with Randy recently than anybody else. Some rangers, who had experienced their own relationship challenges as a result of their jobs, were sympathetic to Lyness and Randy, the oft-quoted line being "Hey, it gets lonely out there." Word on the street, or trail, among other backcountry rangers was, according to Durkee, "Aren't they old enough to know better?" followed by "It's all fun and games till somebody gets hurt." This time it was Lyness, who by now was "in pretty deep emotionally," says Durkee.

On September 16, Randy hiked to Bench Lake to help Lyness pack up and "demobilize" the station for the winter. "After helping Lo take down the station," says Durkee, "Randy told her he couldn't see her anymore because he was going to try to work things out with Judi." According to Durkee, this "devastated" Lyness, and, as the man in the middle, exhausted Durkee even though he wasn't technically involved. He tried to make sense of it. He knew Randy had been in a bad way since his mother's death. He was fairly certain that event had triggered a midlife crisis, which didn't justify his stringing both Judi and Lyness along the last couple of years. Likewise, it "wasn't quite correct to call Lo a victim either," says Durkee. "She was every bit as assertive in pushing the relationship with Randy, and she did know he was married." The only true victim, thought Durkee, was Judi.

CHAPTER TEN

BRING IN THE DOGS

It's hard to feel sorry for a man who's standing on his own weenie.

— *Alden Nash, fall 1995*

Randy didn't tell anybody where he was going on that last patrol. That told me maybe he didn't want to be found.

— *Alden Nash, summer 1996*

SUNDAY, JULY 27: the effort to locate Randy Morgenson had grown to fifty-five persons, including twenty-seven ground searchers, three helicopters, and three dog teams, two of which were dispatched to the cliff area Durkee and Lyness had searched with Cowboy. These dogs weren't trained to follow a specific person's scent; instead, they would alert on any human scent in the area. One dog was also cadaver-trained.

It was the third day of the search, and "word got around that there was a cadaver dog on scene," says Rick Sanger, "and even though that's very common, it really drove the seriousness of the operation straight into my gut. Normally, you can disassociate yourself on a SAR, but hearing the word 'cadaver,' I couldn't help but visualize Randy somewhere out there dead or dying. It really affected me, made me want to do everything possible. In the morning, I'd see some of the searchers taking their time eating breakfast or relaxing with a cup of coffee instead of hustling, and it really pissed me off."

Lo Lyness, on the other hand, was on the trail before 8 A.M., searching the southeast section of Upper Basin, which was northeast of Bench Lake. Her team performed a modified grid search, linking corridors of terrain so as not to miss any part of the area assigned. Even though it was "extremely unlikely" Randy would have taken a route through the steep talus slope above Cardinal Lake, they stuck to the plan and covered that as well. Some fresh tracks between the lake and Taboose Pass were found but were soon discounted as those from another team that had overlapped their search-segment border. Frustrating on the one hand, but effective on the other: better to overlap another segment than to leave a gap.

Laurie Church and Dave Gordon were on the second day of their assignment when they headed north, off the Woods Creek Trail on a cross-country route into Segment M—the Window Peak Lake area. Segment M was almost 5,000 acres of rock, high-alpine tundra, and steep couloirs, with a creek connecting a series of pocket lakes spilling down the canyon over granite shelves. It was one of the southernmost search segments. When the initial team of rangers calculated the Mattson consensus, this area had been given a low 2.2 percent POD based on the fact that Randy had patrolled in this general direction less than a week before he disappeared.

Church and Gordon went up the drainage from the south, using binoculars to examine the high routes into the basin and looking for footprints in the prevalent snowfields. They found nothing to suggest anybody had come through this area—not a footprint or a slide mark or an overturned rock. Unfortunately, twenty-seven searchers covering a dozen other segments had no better luck. On schedule, late-afternoon thunderstorms punctuated the mood of yet another day—Randy's seventh day missing—without a single viable clue.

Drifting gray storm clouds shrouded Marjorie Lake Basin as Lyness returned to the backcountry incident command post at the Bench Lake ranger station. She was exhausted, she'd heard no good news so far, and she understood that, statistically speaking, if a missing person wasn't found on this, the third day of an SAR, "they're either dead or you're never going to find them." With this belief, she walked into the impromptu enclave of the field command post. By this time it was pouring rain and one of the first people to approach her was Special Agent Al DeLaCruz.

Most of the rangers respected DeLaCruz and appreciated or at least understood his difficult job as a wilderness detective during the Morgenson search. The fact that DeLaCruz didn't know Randy very well made his job slightly easier, but since most of the parks' personnel did know and like Randy, he had learned to tread lightly. However, DeLaCruz couldn't discount anything. He vigorously pursued all options. He had already spoken with Judi Morgenson who gave him permission to monitor bank accounts and credit cards, none of which had been accessed by Randy since before the season began. Still, the idea that a person would stage his disappearance in the Sierra backcountry wasn't unprecedented.

In July 1978, David Cunningham was reported overdue from a backpacking trip in Yosemite. The search lasted two and a half weeks and cost the NPS more than $20,000—a significant amount at the time. Search teams were exposed to hundreds of hours of hazardous terrain, including steep snowfields, wild rivers, and sheer rock faces. As stated in the case incident record, "All this effort was unneeded because Cunningham was not lost, but had decided to leave his wife and family for personal reasons" without telling anyone. The mystery was solved only when a friend received a postcard from Cunningham weeks after the search had been called off. He had taken a bus across the country and was lying low in Bangor, Maine.

DeLaCruz's conversation with Lyness began with the same scripted disclaimer as with all his interviews: "I know this is difficult for you—being a friend of Randy's—but I hope you can understand that we're trying to get to the bottom of this. There are some difficult questions

I need to ask you, and I hope you understand it's in the best interest of finding Randy and the safety and well-being of all the searchers involved."

From a distance, the conversation appeared casual but, as the increased volume and body language indicated, "you could tell Lo wasn't taking the questions kindly," remembers Durkee, who was mingling with some other rangers nearby. It didn't take long—DeLaCruz ended the questioning when Lyness broke down, crying. In the interest of privacy, DeLaCruz will not divulge the dialogue between himself and Lyness, but he does acknowledge that it was the most difficult interview he conducted during the SAR and that, due to her emotions, Lyness hadn't offered "any information that was of benefit to the investigation." He was, however, "absolutely certain" she was completely genuine in her distress and honestly had no idea where Randy was.

Indeed, Lyness wrote in her logbook, "By today, there can be no question that Randy's seriously injured or no longer alive." She went on to describe the interview with DeLaCruz: "Got home to face interrogation by 'investigations.' Obviously pursuing a heavy suspicion that RM either left the park or did himself in. Neither option is one any of us who knew him well find possible. He would have to have turned into a person none of us knows to do either of these things. And the insinuation that he might be 'hiding out' is patently absurd. All of those options rate right up there with aliens and a spaceship. When you don't turn up clues, it's always easier to believe the person isn't there. I find the intrusions into Randy's personal life to be jarring and harsh. An unpleasant evening, at best."

ON THE OTHER SIDE of the parks, soaked from a different cloudburst, backcountry ranger Nina Weisman was returning to her station at Bearpaw Meadow. She represented a faction in the parks who struggled through one of the most frustrating, helpless duties in the Morgenson search—not taking part.

Weisman had adopted Randy as her mentor in 1988, the season he'd impressed her "more than words can convey" when he called out over the radio to make sure Robin Ingraham—the climber who'd lost his best friend on the Devil's Crags—was not left alone. Another time, Randy had talked her out of an embarrassing situation when she was a newbie trailhead ranger who got lost her first time off-trail. She hadn't actually been lost, she'd just doubted herself, and it was Randy who came to her rescue via radio. His calm voice had put her immediately at ease. He asked her about her surroundings: What trail had she left from? What did the terrain look like? The trees? Did she hear water? Where was the sun in relation to the peaks? After more than a dozen questions, he reassured her that she was exactly where she thought she was. "He gave me the confidence I needed to go on that day," says Weisman, "but in other ways, the confidence to follow my dream to go for it. Become a backcountry ranger. He nudged me over the edge—off the trails—where I learned to enjoy the real magic of being totally and completely alone in the backcountry."

More than anything Randy had taught her to "pay attention and don't walk too fast. You might miss something."

After having worked eight years at various positions in the park, from toilet scrubber to trailhead ranger to bear management specialist, Weisman had finally earned her ultimate assignment in 1996: a backcountry ranger with her own station. Randy was one of the first people she'd shared the thrilling news with, and she looked forward to many more seasons working with Randy, whom she described as "one of the kindest souls to ever walk the trails in this park."

Now he was missing, and she was angry because she had not been chosen to take part in the SAR, despite her knowledge of the search area.

Weisman had spent the day cleaning fire pits around Mehrten Creek; on the way back, the rainstorm hit. "I was a mile from home after a depressing 12-mile day," she says. "I was listening to the radio traffic about the search and was really upset. I couldn't help thinking something bad had happened to Randy, so I was just plodding through the rain, getting madder and madder that I wasn't invited to help in the search—kind of having one of those internal conversations—and I tripped on some rocks, fell forward, and ripped my knee open.

"I was close to home, so I wrapped a bandana around it, and there was blood and soot from the fires and it was streaming down my leg from the rain. The pain started throbbing, and I thought, 'Geez, I was depressed and mad and it made me not pay attention, just for a second, and look what happened.'

"I knew from training that Randy was having a really bad summer and was depressed about stuff, and it hit me that he was probably not super careful, not on his toes, not in that place you have to be mentally out here, and he might have just tripped, except in a very bad spot. I was on a trail and just hit rocks. Randy, I knew, was somewhere off-trail. I got more and more depressed—it was hard for those of us who wanted to help but had to listen to the radio and watch the helicopters flying all over the place. That night I couldn't help but think Randy was out there seriously injured.

"I can't tell you how many times I considered abandoning my post to go join in the search. That's how worried I was. But then I realized I wasn't the only one. There were other rangers who wanted to take part, but there was still a park full of people. There were still bears stealing food, and injuries to attend to, and permits to check. My presence was needed here at Bearpaw. I told myself that every single night of the search."

AFTER THE DISAPPOINTING results of day three of Randy's SAR, the overhead team members at the Cedar Grove fire station command post racked their brains for alternative search methods. While conversing with one of the California Highway Patrol officers whose unit was donating personnel for the cause, Dave Ashe was told that the Army Air National Guard in Reno possessed a special asset that could prove helpful—a night helicopter equipped with forward-looking infrared (FLIR), which detected body heat. Ashe, who had already used California's Office of Emergency Services (OES) to disseminate their needs to the SAR community, again approached his OES contact with the hope of gaining access to the military's state-of-the-art technology. The initial problem was that the SAR was in California, and Reno is in Nevada. But since the California Guard's ship was not available, the OES contacted Nevada's Department of Emergency Management, which authorized the mission.

Chief Warrant Officer III Bob Bagnato and Chief Warrant Officer II Darren Chrisman got the call a little before 9 P.M. on Saturday, July 27. Their OH-58 (Bell Jet Ranger) helicopter landed at Cedar Grove three hours later.

At 0100, Bagnato and Chrisman, dubbed "Recon 71," were high above the search area and getting a bearing on the boundaries of the assignment by referring to their GPS and visual landmarks—in this case, massive granite ridgelines that glowed green through their night-vision goggles.

Even though many military pilots had bootlegged low-level flights into Kings Canyon National Park, this was apparently the first authorized night flight. Coffman's admonition to "try and focus on the higher, more rugged terrain that isn't easily accessible on foot" was something Recon 71 had agreed to, "if conditions allow." Now some of the higher ridgelines were shrouded by cloud cover, but the wind was mercifully light, making navigation a little less wicked.

The biggest concern with any search, but especially a search at this altitude, was maintaining enough speed and altitude to allow pilots to fly out of any potential mechanical failures while sticking to their search-and-rescue mantra: "slow and low." That meant 20 to 40 knots; any faster and they'd lose their edge as an effective search tool. Bagnato and Chrisman were dedicated night fliers, so their missions were often poised between the operational periods of searches. Normally, there is very little small talk during briefing and often they don't even know whom they're searching for. "Just a warm body," says Chrisman. But this time they knew it was a ranger, and that it was his seventh night out there alone.

"That just stuck out for me—lucky 7," he says. "I started that night out with real high hopes that we'd find him."

Between the two, Bagnato had more mountain flying hours, so he was the designated pilot in the right-hand seat while Chrisman operated the FLIR camera in the left-hand seat. Standard strategy was to fly one "high recon," about 1,000 feet above ground level, to take a look at the search area and identify any points of interest: trails, creeks, anywhere a person would likely gravitate toward. Then they would make a lower pass over these points of interest, which could include game trails that were difficult to see on the ground but with night-vision technology looked "like a sidewalk."

Once these two passes were made, a low-level grid search would follow strict GPS coordinates, first east to west and then north to south. The two systems, night vision and thermal, work well together: what one can't see, the other can. In densely vegetated or wooded areas, the same lines are flown both directions to better see into the varying degrees of cover.

The pilots conversed almost constantly, Chrisman focusing 70 percent of his attention on the FLIR screen in front of him and 30 percent out the window. Meanwhile, Bagnato was focused 100 percent out the window, scanning left to right for any potential light source. A lit cigarette a mile away was all it would take to send them off their grid pattern to investigate, but only after marking the exact location where they diverted.

The parks' helicopters weren't equipped to fly at night, so when the search teams camped at the Bench Lake ranger station heard the distinct, deep whoop-whoop-whoop of the Jet Ranger blades cutting through the cold air at two in the morning, it was a bit disconcerting. Nobody had alerted them about the night operation, which Durkee summed up with two words: "voodoo spooky."

The moon was nearly full and the surrounding granite basin was lit in ghostly Sierra light. When the helicopter passed overhead, treetops rustled. Then its darkened shape, clearly outlined against the stars, faded away to the north. Despite the searchers' exhausted bodies and minds, rotors beating wind invoked an adrenaline rush that kept some awake for the rest of the night.

Around 3 A.M., Bagnato said, "I have a campfire." Chrisman confirmed, and Bagnato descended into the depth of what was LeConte

Canyon. Well off any trail was a figure, apparently sleeping beside a fire. With night-vision technology from two miles away, the small, smoldering fire had looked like "a circus, or the Vegas Strip." Going low, Chrisman used the finger touch pad on his controller to zoom the thermal camera in on the person, then videotaped the scene, simultaneously recording the exact GPS coordinates. The coordinates were relayed back to the incident command post, where someone would be assigned to investigate on foot since landing in the narrow gorge was not possible.

No other solo warm bodies besides those of the resident wildlife revealed themselves that night, but Chrisman had watched a large buck urinate (through the thermal imaging, the ground around the deer's hind legs appeared "white-hot" as the puddle spread). "Serious wild kingdom footage," says Chrisman. "It was the capper for the evening." Both pilots felt confident that the solo hiker huddled next to the fire was the missing ranger.

By sunrise, everyone at the Bench Lake ranger station was aware that the "voodoo" ship had been a military loaner. Shortly thereafter, it was confirmed that this technologically advanced, multimillion-dollar eye in the sky had identified one potential warm body. The person was discovered not to be Randy.

ON THE MORNING OF JULY 28—day four of the search—various volunteer groups and state and federal agencies had joined the Sequoia and Kings Canyon rangers, upping the total number searching for Randy to sixty-three, including thirty-four ground searchers, five helicopters, and four dog teams.

At the Bench Lake staging area, Rick Sanger learned his assignment was to sweep the high ridges northeast of Mather Pass, a locale he found "unbelievable."

"Why are they assigning me these high, impassable ridges that any meadow stroller like Randy would just shake his head at?" he thought at the time. Sanger's incorrect assumption was that the aging naturalist would have avoided such vertical granite mazes. In reality, Randy loved high places and had finessed his way up into many of the parks' zones generally reserved for the resident mountain sheep.

Moments later, Sanger was introduced to the dog handler who would be accompanying him, an official-looking officer from the Department of Fish and Game. Her camouflage pants and 9 mm pistol holstered at her side gave her a Special Forces mystique. Sanger's immediate reaction was "I wouldn't want to get caught poaching by this person." Then he saw Kodiak, her rottweiler search dog.

The three climbed into a waiting helicopter and, as Sanger donned a helmet, Kodiak started growling. The handler told Sanger that she suspected the aggressive behavior was because his green Park Service uniform and helmet resembled the "bite suit" Kodiak had been trained with. Sanger asked if the handler had a "scent item" from Randy's belongings, but she explained that Kodiak didn't track that way. Rather, he was trained to follow "disturbed areas," wrote Sanger in his logbook. Regardless, Sanger delayed the flight while Durkee brought them one of Randy's shoes.

They landed and, after the helicopter's departure rotor wash settled, Sanger spread a map on the ground to go over their search route. As he kneeled, Kodiak lunged from three feet away and sank his teeth into Sanger's hand.

Shaking, Sanger walked to a nearby stream to wash his hand and watched blood swirl into the current from two puncture wounds. Taking deep breaths, he tried to convince himself this day wouldn't be a total waste of time, even though he was pretty certain Randy hadn't been packing any animal gallbladders, the one scent item Kodiak had been trained on.

"For Randy," says Sanger, "I composed myself and returned to the handler, who was extremely apologetic and, taking the role of doggie psychiatrist, guessed that Kodiak might be 'feeling threatened.'" Sanger wrote in his logbook that night: "I could relate."

As predicted, their search didn't provide any clues, but it did serve an important purpose by closing another gap in the search area. That night, Sanger recounted the dramatic day to Durkee, stating how ironic it was that this was the first patrol he'd been on without his duty weapon. Durkee nodded, saying, "It's a good thing. She looked to be a quicker draw."

For Kodiak, it was his last day on this search. No room for dogs without good manners.

BY THE END OF THE FOURTH DAY of the SAR—the eighth since Randy had last made contact—many of the rangers were feeling its mental effects. At the request of veteran rangers, Dave Ashe telephoned Alden Nash at his home in Bishop, California.

Most of the rangers searching for Randy had worked under Nash during his tenure as Sierra Crest subdistrict ranger from the mid-1970s until his retirement in 1994. They felt they needed the emotional, even fatherly, support Nash had provided them as their supervisor. In addition, Nash had hiked with Randy more than anybody in the high country and would be a valuable asset.

Nash graciously refused the request to join the SAR, rationalizing that it wasn't worth risking his life after Randy's recent admissions. In the first place, Nash thought it highly possible that he had left the mountains. A phone call from Ashe early in the search asking if Randy had been in contact with him confirmed that this line of thinking wasn't solely his own.

But if Randy was indeed injured somewhere, Nash reasoned that whatever he had gotten himself into was a result of the choices he had made in his life. Nash felt that Randy's mind likely hadn't been in the right place, that he'd been severely depressed and had made a mistake or even done himself in.

During his thirty years with the Park Service, Nash believed that when he put on the uniform, he had an obligation to uphold what it stands for. "For me, that sense of duty overflowed to all aspects of my life—meaning my family," he says. "I always thought Randy felt the same way, but he was living a lie. Learning that was a big disappointment. It was like a kid finding out the truth about Santa Claus. Randy had been, in my eyes, the epitome of ethics and morals, and here, all of a sudden, he was human. I knew he was ashamed of it or he wouldn't have kept it from me for three years."

Nash's sense of duty to his family after his retirement was to be around for his grandchildren. He felt fortunate that he'd come through all those years of service for the most part unscathed. He also didn't like helicopters. "Every time you set foot inside one," he says, "you're risking your life." He had in fact experienced multiple close calls on helicopter flights into the mountains. On one such occasion, he watched a helicopter he'd just stepped off 30 seconds before crash-land because of an engine malfunction. "Helicopters fly," he explains, "by beating the wind into submission.

"Then there's gravity. Gravity plus granite equals a god-awful mess."

Upon getting off the phone with Ashe, Nash's resentment toward Randy grew. "Randy knew the dangers involved with an SAR," he says, "and I thought he damn well better be hurt, because if he left the mountains and somebody else gets it, there might be some serious repercussions. Some people might consider that some degree of murder."

Even with his suspicions and anger, Nash was quietly concerned about Randy's well-being. Not joining the SAR was still a difficult decision.

On that same day, DeLaCruz's investigation team reviewed Randy's 1995 LeConte Canyon logbook, homing in on the incident with Packer Tom and the altercation with Doug Mantle. Further, DeLaCruz was made aware that Mantle had written a bitter account of the citation that Randy had given him for improper food storage and an unattended camp. The article was published in the issue of *Sierra Echo* that came out seven months before Randy's disappearance. The irony was that Mantle *had* properly stored his food in bear-proof canisters. His companions were the culprits, but Randy had cited the person whose name was on the permit—no doubt with some sense of retribution when he found it to be Mantle. In the end, Mantle paid a nominal $85 fine, but he also missed two days of work and had to drive 440 miles round-trip to court, plus pay room-and-board expenses. Mantle's article also conveyed that "nobody ever heard of a rule prohibiting leaving gear unattended" and if there was such a rule, the NPS had failed to advise the public about it.

DeLaCruz wasn't interested in passing judgment on Mantle's innocence or guilt regarding the citation. He was more interested in clues regarding motive: Was Mantle exceptionally angry with Randy? In the *Echo* article Mantle wrote, "Beware. There is one lurking in the backcountry, waiting to do you in. You might not spot him in daylight (if you're lucky) but with stealth he can rise up and ruin your whole outing. Well, or at least create a new outing for you. I refer to Ranger Randy." He concluded with "The National Park Service has embittered this customer with this idiocy. The obvious result is that cooperation will be grudging. . . . Out-of-control bears are a lot more palatable at times than rangers."

DeLaCruz considered both Packer Tom and Mantle people worth checking up on.

THE EFFORT TO LOCATE RANDY continued to grow on July 29, the fifth day of the search: sixty-nine personnel, including thirty-seven ground searchers, four helicopters, and five dog teams.

Bob Kenan spent the day at Simpson Meadow ranger station, one of his favorite areas in the park and the last place he'd seen Randy in the backcountry. His assignment was to interview hikers who passed through and maintain a presence in case Randy came to the station. Simpson Meadow was at a confluence of several backcountry routes.

At Simpson Meadow, Kenan was overcome by the magnitude of what was going on. "The SAR at that point was just this amazingly powerful and emotional event that I will never forget for the rest of my life," he says. "It encompassed the entire park. Frontcountry personnel, backcountry rangers, researchers, trail crew, scientists: we were all, in our own way, doing everything possible. It became a desperate search to find Randy and save him if that was at all possible. That was our goal and our mission. It was not a search for a body; it was a search to try and save Randy."

Despite nearly twenty years working together as backcountry rang-

ers, Randy and Kenan had not been close. "We had times and encounters within the mountains that were positive," says Kenan. "But that
wasn't always the case. The season before Randy went missing, he'd
hiked from his LeConte duty station to pay Kenan a visit at Simpson
Meadow. It was one of the few times they'd spent an evening together
socially since 1978, when the two had patrolled together near Rock
Creek. Another time they had put up some bear poles in the Kearsarge
Basin. Otherwise, their meetings were relegated to training, search-
and-rescue operations, and the occasional chance meeting when their
patrol areas overlapped. During these encounters, Randy wasn't always
the friendliest coworker, though Kenan won't elaborate other than to
say, "It was a sort of cold shoulder."

So, after more than a decade of occasionally unfriendly encounters, Randy surprised Kenan by hiking over to Simpson Meadow to
see him. As the two looked out over one of the High Sierra's wildest
meadows, Randy said out of the blue, "I've been an ass to you, Bob. I
understand now how I've treated you, and I don't really know why I
haven't been so friendly to you over the years."

"The admission was nothing less than an apology," says Kenan. "It
was an extremely brave and honorable thing to do, because I *had* taken
it personally over the years. We had dinner and just reveled in our
mutual appreciation of the power and beauty that surrounded us. Our
rocky history wasn't totally forgotten, but it was filed. It really cleared
some of the tension and negative vibes that we had between us over the
years. It was a new friendship and a new start."

Now, with Randy missing, Kenan had a difficult time not wondering if Randy's apology had also served to clear his conscience. He
shook the thought out of his mind. Randy was out there somewhere,
alive—and they would find him.

But at day's end—the eighth day since anybody had heard from
Randy—the only significant event was when two frontcountry rangers, Claudette and Ralph Moore, were chased off State Peak by a lightning storm. They'd been climbing summits to check the register boxes
just in case Randy had signed in. Searchers had found: a moldy shirt

from the season before; the same size 9 boot print on Cartridge Pass (searchers' paths were beginning to cross); a yellow piece of paper (found to be a searcher's); and toilet paper in a Ziploc bag.

Nothing was linked to Randy.

ONE WORD CAME UP time and again to describe how the backcountry rangers felt at the peak of the search: numb. It was, according to one ranger, "a circus in a *Twilight Zone* episode," an overwhelming buzz of activity coupled with the odd sensation that they were both spectators and actors on a bizarre wilderness stage. From a personnel standpoint, almost one hundred searchers were focused on finding Randy on July 30—fifty on the ground, including a dozen dog teams, and four helicopters in the air.

By now, every segment had been searched. Unless Randy was purposely hiding out or he'd left the mountains, there was a strong possibility of stumbling upon his body, something the searchers had to steel themselves against as they were inserted into areas already searched. There was even a code word to use over the radio if they found Randy's body, so local media didn't intercept the message before next of kin was notified.

Judi Morgenson, who was getting daily phone updates from both Chief Ranger Debbie Bird and Special Agent Al DeLaCruz, had become convinced something was truly wrong, in part because of a dream she'd had—not once, but twice.

"I was driving down this mountain road and came to a clearing and there was this lake with granite along the shoreline," she says. "Big trees were hanging over the surface kind of like the bayou, and I looked into the water and it was crystal clear and there was a man with a backpack floating at the bottom of the lake."

It was vivid—a clear vision that Judi couldn't discount. She asked Debbie Bird if they had checked the lakes in the area. Bird responded that they had been looking "everywhere." They hadn't brought in divers, as there were literally hundreds of lakes in the search area, but shorelines were searched, as were major waterways. No telltale footprints led them to believe that Randy had fallen into the water, and no dog teams had expressed interest in such locations—at least not before a dog named Seeker started searching a low-priority basin at the southern end of Segment M.

Like thousands of search-dog handlers nationwide, Linda Lowry had a "real" job; in her spare time, she volunteered for the Contra Costa County search-and-rescue team and was a member of the California Rescue Dog Association (CARDA). Lowry and her dog, Seeker, represented the fifth or sixth wave of search-dog teams to descend upon Kings Canyon. Most had lasted only one or two days before the rough terrain and high altitude disabled them.

Lowry had been dispatched the afternoon before, driven five hours to Cedar Grove, got settled at 3 A.M., woke at 5:30 A.M., was briefed at 6 A.M., and was waiting for a helicopter at 9:30 A.M. From other dog handlers who'd been in the field she learned that this search was "very personal" and that "They're pushing it hard." This concerned her.

As a dog-team handler, it was her job to run the search. But since the terrain was so remote and dangerous, none of the dog teams were allowed to enter the search area without rangers, who, in essence, acted as guides. Apparently, some of the dog handlers who had been teamed up with close friends of Randy had difficulty "controlling" the rangers. "They had their own ideas of where Randy would be, but you can't allow those hunches when you're working with a dog," says Lowry. "You have to put the dog to the advantage of the wind and the terrain and stick to the system you're using, which doesn't always seem like the best plan, especially if you're a ranger and you know the territory backwards and forwards." The rangers knew Randy wouldn't be out in a place such as a flat, open meadow, so why waste your time with it? You can see clearly that he isn't there. "But as a dog handler," says Lowry, "you have to cover all of your area. Otherwise you don't get the maximum benefit out of your dog."

Lowry was relieved when she was introduced to Rick Sanger, who thanked her for being there and immediately made her feel at ease. She didn't know it, but Sanger was equally relieved when he met Seeker—a giant schnauzer that he perceived was a friendly dog. Their assignment was the Window Peak Lake drainage, the same segment searched by Dave Gordon and Laurie Church on the third day of the SAR. They would meet up with Bob Kenan's team, already working its way into the drainage on foot, carefully scouring some of the dangerous cols that led from the adjacent Arrow Peak Basin.

The Window Peak drainage is an experts-only "detour" from the John Muir Trail. On the grid, it would be like cutting a corner, or taking the shorter, bumpier scenic route instead of the longer, faster interstate highway. It's rough country—some of the roughest in the parks—though you can still travel without a rope if you know how to weave around the cliffs that abound in the area. Though the mileage is technically shorter than taking the longer trail route, the time spent navigating and grunting through no-man's-land would likely double a hiker's time, if he found his way at all. Even rangers get turned back on occasion, for the routes into this high valley are often packed with snow and ice the entire summer. For this reason there are seasons when the Window Peak drainage sees barely a footprint.

"Trail miles and cross-country miles are completely different animals," says Bob Kenan. "And air miles don't mean a thing. If you think in terms of 'how the crow flies' in the Sierra, you'll end up in a world of hurt.

"Two or three hundred yards across a talus field might take an hour, where you can jog that same distance on a designated trail in less than five minutes. Cross-country routes can eat you up so bad you feel like kissing the ground when you hit a trail. It's not too hard to find places that haven't seen a footprint in three or four seasons. It takes effort, but there's plenty here in the park. Plenty."

The Window Peak drainage is enclosed by an assortment of high (up to 13,000-foot) peaks, vertical cliffs, and high-angle glacial moraines, littered with Volkswagen-size boulders that are known to shift and occasionally tumble down with the slightest disturbance. But even though the drainage was deemed a low-probability area, it still had to be searched—twice. "Low probability of area doesn't mean it's a low probability for injury," explains Sanger.

One of two or three plausible routes into the drainage from the north (off the Bench Lake Trail), Explorer Col is steep, made up almost entirely of loose rock, and usually filled with snow late into the season. An ice ax is often mandatory to negotiate the final crux.

But once over that col, the Window Peak drainage pours out below like a land that time forgot—bubbling creeks, wildflowers, and sapphire blue mountain lakes seem to drip down the landscape like eye candy from above. At the bottom, the creeks converge into a narrow chasm, which eventually enters Window Peak Lake, nestled beneath two dominant peaks to the west: Window Peak (named for a hole in the granite that makes a "window" on a ridgeline near the summit) and Pyramid Peak (named for its shape). It was, using Randy's description, "rich country."

Upon landing at the upper end of the basin, Lowry put the orange shabrack search jacket on Seeker, and the dog's playful demeanor evaporated. Time to go to work.

Bob Kenan and Charlie Shelz, who had been scouring the basin's western slope, rendezvoused with Sanger and Lowry. The scale of this basin was such that two search teams on opposite slopes would appear to each other as dots smaller than ants. Movement was often the only visual clue alerting searchers to one another in this vast mountainous terrain. After a brief discussion, the two teams decided on a strategy that would both thoroughly cover the segment and take advantage of the wind that funneled up the basin with an easterly flow, keeping Seeker in the right zone to smell anything down-canyon. Shelz and Kenan would continue down the western slope, while Sanger, Lowry, and Seeker would cover the eastern slope above where the helicopter had landed, then sweep back down the canyon to the center of the drainage.

Seeker was a certified air-scent search dog that had been introduced to cadaver scent, but not enough to build an alert around. CARDA requires that each dog pass a 20-point skills checklist in preparation for a certification test, during which Seeker found two mock victims hidden in 160 acres of wilderness in under four hours. Likewise, Lowry herself had to pass a laundry list of requirements. Lowry had also trained

Seeker in tracking, although the dog wasn't certified. "Tracking is always a skill they can fall back on if the day is extremely hot or a scent is evaporating," she says.

If Seeker smelled human scent while on a search, she would follow it briefly to confirm and then return to Lowry, who had a tennis ball attached to her belt with Velcro. "She loves to hear that ball rip off my belt," says Lowry. "It's a big game to her, but it's also her alert—all dogs have their own distinct alert."

Generally, a search dog is put to the roughest terrain in an area first, but in the Window Peak drainage everything was difficult—talus fields, steep slopes, snow. Normally, in rocky terrain, Seeker would have been "booted up," but the sporadic snowfields proved too slippery for her neoprene booties. For safety, Lowry kept them off.

As they began the search, Lowry stayed toward the center of the basin, where, Sanger warned her, there was a creek. To Lowry it all looked like a snowfield. With Seeker in search mode, Lowry was so pumped on adrenaline, she didn't even feel her lack of sleep. She did, however, feel the 11,000 feet of altitude, starting with a headache that quickly traveled to her stomach.

On the way down the basin, Sanger fanned off to the east and Lowry gravitated toward a distinct spine near the basin's center—a classic route a person might travel because it provided some elevation to visually scout the terrain ahead. To the west of the spine and the creek was snow; to the east, broken granite. The spine itself was a mix of snow, talus, broken granite—the type of terrain you can't take your eyes off of for a second.

A strong draft blew up the basin from the south and west—perfect conditions for Seeker's nose. Moving down the basin in a serpentine pattern, Seeker would periodically come back up the spine to keep within 50 or 75 feet of Lowry, who had vomited numerous times but refused to abandon the search. "Altitude sickness can be horrible," she says, "but you have to remember what shape the victim might be in and that you might be their last hope. As long as I'm not a liability, I'm not stopping a search."

Midway down the spine, its western slope steepened, the wind increased, and the snow took on a different consistency, with slippery sections of ice, some of which had a sheen to it, according to Lowry. Here, Seeker suddenly deviated from her search pattern and angled down and off the spine. Fifty feet below, the dog broke through the surface of what appeared to be a frozen lake.

Lowry watched in horror as Seeker dropped into the dark water. Lowry couldn't move quickly for fear that she'd slide down the incline and end up in the lake as well. Nobody else was close enough to assist, so she picked her way down the slope, calling out "Allez! Allez!" which in French means "go," but which Seeker knew as "come!"

Lowry was certain she was about to witness her dog's death. The hole in the rotting ice increased in diameter to 10 feet as Seeker swam from edge to edge, struggling frantically to find purchase. Finally, she lurched up and out of the water, her momentum sliding her toward thicker ice, where she got her footing and scrambled up the slope to Lowry.

Seeker lay panting, shivering, and bleeding. Warming her dog as best she could with a jacket, Lowry then took a reading on her GPS and scribbled the coordinates down: "4084.5 North/370.9 East."

"Seeker bolting off like that was completely uncharacteristic for her while in a search pattern," says Lowry. "She did not alert traditionally, but she never really had the chance to come back and rip the ball off my waist. I interpreted her behavior, certain that she'd caught scent of something human, and wanted to mark the location."

Sanger's radio sprang to life when he was about 100 yards from Lowry's location. "Seeker's got a split-open foot," she told him.

Sanger's first thought was an image of Randy's bloody foot in Seeker's mouth before he realized that Lowry was talking about the dog's paw. He worked his way down parallel to Lowry and Seeker, who were picking the easiest, most benign route possible for bandaged paws.

They headed toward Window Peak Lake, a prearranged helicopter landing zone. Along the way, Sanger continued his search technique, zigging and zagging down the broken scree. He tended toward the creek, where he could hear water cascading in a falls. Snow dominated the center of the drainage, and he kept an eye out for footprints. Near the falls, he maneuvered along the edge of a cliff. Leaning out on an outcropping of rock, he poked his head under an ice bridge that had formed across the creek. He was hit by a cold burst of air from the turbulent, rushing water forced through a dark, cavernous snow tunnel. It didn't look like anyplace Randy would be if he were alive, so Sanger continued down the drainage, meeting up with Lowry in time to see her off in a military helicopter around 7 P.M.

Kenan radioed Sanger to let him know his team was going to camp higher in the basin. They were taking it slow, really dissecting the terrain and not leaving a stone unturned. Sanger, who, unlike most of the searchers, was not yet entertaining the idea that Randy could be dead, thought that if Randy wasn't injured, he'd gone south, maybe hitched a ride with some trucker on U.S. Highway 395 to a border town, where he'd slipped into Mexico. "The sky," after all, "is the limit." That's what Randy had told Sanger only weeks earlier.

Maybe he was on his way to Argentina, heading for some unspoiled wilderness in Patagonia. Right now, he might be holed up in Tierra del Fuego, sipping beer with a pretty Argentinian, having set up his fellow backcountry rangers with mega hours of overtime and hazardous-duty pay while they searched for a ghost. He could have predated that note and been out of the mountains in the dark before the hermit thrush even woke up. This "gone south" theory was not something that Sanger truly believed, but it seemed a happy vision to dwell on while he set up his camp in a stand of lodgepoles near Window Peak Lake's northern inlet.

Around 8 P.M., he radioed the incident command post and told the dispatcher, "If you'd like to put another dog team out of commission, I'll be waiting." Not long after, he was told that a dog team would be flown in to meet him the following morning.

Sitting in the gravel on his sleeping pad, he boiled water and watched the Window and Pyramid Peak ridge meld from gray into black, then slowly back to a silvery opalescence with the rising moon.

The color of the granite, the diminishing rustle of pine needles, even the tone of the lapping waters on the lake's shore took on the personality of night. The wind had calmed, the birds had gone to nest, and a higher degree of silence surrounded him. He crawled into his tent, pulled off his boots, and, once in his bag, focused on the sound of the waterfalls in the gorge as they cascaded toward him, lulling him to sleep.

By the evening of July 30, the rangers were experiencing varying emotions. A classic mind-set chronology during a search-and-rescue operation is marked by hope at the beginning of a search, followed by doubt as searchers question the areas they are searching, and then frustration or desperation when a person isn't located. From there, searchers often feel a numbness—a sort of stall point where they just follow along dutifully, not sure how to react to their emotions other than to move forward. Some searchers resort to denial and won't allow themselves any premonition except the best outcomes, while others, the realists, tend toward statistics and prepare themselves for the worst.

The strain of Randy's disappearance had reached far beyond the parks' borders, having been broadcast by television, radio, and newspaper reporters and the NPS Morning Report, a daily update on incidents in the national parks. Most rangers across the nation knew one of their own was missing in the rugged High Sierra. For Lo Lyness, the search was over. She had left the mountains to attend to a dental emergency and was home in Bishop, California. "I could have been another body roaming around, but it would have been a miracle if he'd been found at that point," she says. "Statistics were against it." That was the realist, the stoic ranger. The friend and former lover was grieving and in pain. For that side of Lyness, it was "easier to read about the jet that blew apart in New York than to think about the search," she wrote in her logbook that evening.

Durkee was melancholy, but driven by both hope and urgency because maybe, just maybe, Randy was in some nook or cranny of Kings Canyon—too weak to signal, too injured to move—waiting for help. Durkee had been flown from the backcountry to Cedar Grove with Randy's gear. At the incident command post, he discovered some park administrators reading Randy's personal diary. He understood why DeLaCruz or perhaps Coffman might glean pertinent knowledge from the diary, but for anyone else it was, in his words, "an invasion of Randy's privacy—it was nobody's business." It pissed Durkee off. Once DeLaCruz got his hands back on the journal, "to his credit," says Durkee, "he locked it up." A copy, however, was en route via overnight carrier to a profiler with the California Department of Justice, who would provide an analysis as soon as possible. For the search effort, that wouldn't be soon enough.

Sanger's emotions on July 30 can best be described by a dream he had that night as he slept on the shores of Window Peak Lake. He was jolted awake from a deep sleep by a vision of Randy stumbling into his camp and collapsing on his tent. He interpreted the dream as a message "not to give up."

Sandy Graban and George Durkee had volunteered to box up Randy's personal items at the Bench Lake outpost earlier that day. The cardboard boxes with Randy's writing, his books, clothing, and camera equipment—everything was getting shipped out of the backcountry. The obvious message was that Randy wasn't coming back.

While the two rangers were inside the claustrophobic tent walls, the radio came to life, a happy voice interrupting the somber moment like a clown at a funeral: "Hey there! Everybody okay up there?" Sandy Graban's soft-spoken demeanor flew out the mosquito mesh window. "We're fine," she said. "We're just fucking fine."

CHAPTER ELEVEN

THE WILDERNESS WITHIN

I wish I knew where I was going. Doomed to be "carried
of the spirit into the wilderness," I suppose.

—John Muir, 1887

The harshness of noisy motorized brightly lit civilization,
upon first emerging from the mountains, is a bittersweet experience. Better to feel its crudity than not.

—Randy Morgenson, McClure Meadow, date unknown

ONCE RANDY WAS OUT of the mountains in the fall of 1995, the back-
to-civilization decompression process began. It had been more than a
year since he'd visited Stuart Scofield, so he made the long drive from
Sequoia and Kings Canyon through Yosemite and over Tioga Pass to
the town of Lee Vining, where Scofield had moved.

Scofield, who was happily married and expanding his workshop
business, missed the days when he and Randy spoke for hours about
the mountains and photography. On this visit, he tried to broach those
subjects, but after a few tries, he realized Randy wanted to talk only
about his problems. The thing that was impressive to Scofield, "and not
impressive in a good way," he clarifies, was how Randy had changed.
"He had been such a curmudgeon before, but then, all of a sudden,

he was vulnerable—mixed up, stressed out, *vulnerable*—and that just blew me away. It scared me. It really did. He had lost that aura of invincibility."

The conversation was exhausting. Randy was "absolutely consumed by turmoil" and looking for ways to fix his "problem" so he could continue his relationship with Judi. He probed Scofield for advice on how to get past his attraction to Lyness and go forward with his life with Judi, because he loved his wife and said that was "the right thing to do."

"It's sort of what happens to people if they conceptually believe in fidelity but then are incapable of it," says Scofield. "It sets up a conflict, and if you're like Randy, it eats at you. On the outside, he tried to remain strong and desirable, but . . . he felt guilty because he knew that Judi had always been a good wife, and faithful, but he was conflicted because Lo represented things that Judi was not."

The one word Randy used over and over again was "complicated." His situation was "complicated."

But Scofield knew it was worse than complicated. He described Randy as being in a tailspin and was worried that he might be suicidal, "mainly because there had been such a departure from his normal, confident persona. That alone would have made anybody who knew him well worry about the possibility."

Randy had grown accustomed to a mutually distant relationship with Judi. But upon his return to Sedona, "he wanted to talk," says Judi, "and I knew he'd been with Lo that summer and I didn't really want to listen at this point. I was pretty fed up with him and felt betrayed, taken advantage of, stupid. You name it, I'd felt it. Not to mention, he was so self-absorbed, he didn't ask about my mother."

Judi called him on that, and Randy apologized. He admitted that he'd been selfish and a horrible husband beyond this one instance. "He wanted me to forgive him," says Judi, "but I thought that was just to ease his conscience—he also said it was absolutely over with Lo." Randy begged Judi to believe him, but she'd heard that line before.

Once they settled into what Judi described as their "comfortably distant" relationship under the same roof, Randy told Judi about the two encounters he'd had that summer in the backcountry where he'd felt "threatened." That struck Judi as odd, because they had really shaken him up, something that had never happened before. As Randy explained the encounters, he told Judi that he felt uncertain about himself, wondered if he'd become a different person and had come off as "an asshole" to Packer Tom and Doug Mantle. He told her that after that, he'd become nervous when he approached people because he didn't want to be misinterpreted. "It was a time when Randy wasn't feeling really secure in himself, either," says Judi. "So I told him, unless he had turned into a totally different person out there, that he had just met a couple of jerks. Randy wasn't out there to enforce laws. He was there to try and get people to understand. He got along with everybody— the climbers, fishermen, everybody. He just wanted to make sure they weren't injuring the wilderness, screwing up his mountains."

Judi apparently was right. Years later, one of the two climbers who had witnessed the Doug Mantle altercation, Barbara Sholle, confirmed this: "Randy had merely asked to see our permit and was making friendly small talk. Doug's response was incredibly rude and uncalled for."

For his part, Mantle—a Sierra Club leader for over three decades— maintains that Randy baited him with questions. However, he regrets not keeping his cool during the "altercation" in Dusy Basin. "It is indeed a matter of the moment for me," says Mantle. "I can be sweet as pie, but on a bad day I've gone too far."

EVEN THOUGH JUDI had expressed sympathy for Randy's encounters in the backcountry, she didn't forgive him. He had to earn her trust, and that, she told him, would take time. Judi knew Lyness had moved to Bishop, and if Randy was trying to make things right, he'd better make a few sacrifices in his life. That winter, he opted not to do the snow surveys out of Bishop into the eastern Sierra.

Not that it would have mattered. That unexpected breakup at the Bench Lake ranger station had been the last straw for Lyness. She would always care for Randy, and she loved almost everything about him in the mountains. "Randy just was the Sierra to me," she says. "He was so observant of every little thing around him . . . and he so delighted in all the small things—the rare plant in Grouse Meadow, the duck on the lake in the fall. He honored everything about the Sierra backcountry. He saw so much more than most people ever will." The times they shared together had made her mind wander, probably too far down a romantic mountain pathway, where she'd allowed herself to entertain thoughts of a life with Randy once he left Judi. But she, like Judi, had learned a few things, so when a new man entered her life, she didn't resist. That winter Lyness met her soul mate and his name wasn't Randy Morgenson.

Randy had no idea Lyness had begun seeing someone else. She was another on the list of people he'd alienated, including some of his best friends: George Durkee for one, and Alden Nash, who had been his biggest fan and proponent for many years. When Nash found out about the affair, and how Randy had purposely kept it from him, he wasn't pleased. Randy realized he needed to work some things out on his own. Judi thought this was a good idea.

Over the winter Judi attended a workshop in Sedona entitled "The Artist's Way." It was a course she thought would help her break some creative blocks she was dealing with. She thought the exercises she learned might help Randy as well. "It became a family thing," she says, referring to the free-form writing technique described in Julia Cameron's book *The Artist's Way: A Spiritual Path to Higher Creativity*. The book outlined "a course in discovering and recovering your creative self." This was something Randy knew he needed. The backbone of *The Artist's Way* was stream-of-consciousness writing, a process called the Morning Pages in which a writer simply gets down on paper three pages of whatever comes to mind first thing each morning. For both Judi and Randy, there was plenty swirling around in the dark recesses of their minds, and the Morning Pages became the perfect place to get these thoughts off their chests. These innermost thoughts—letters, so to speak, to and from themselves—were not meant for anybody else to read.

Randy also continued to read a book he had been studying for a couple of years called *Iron John: A Book About Men*. The author, Robert Bly, had written an insightful, complicated best-seller that was highly praised by some and bashed by others for its study of modern man's role in family, life, and the universe. You either love *Iron John* or you hate it. When Randy described it, he said, "It spoke to me." Especially the initial chapter, "The Pillow and the Key." Written as a fairy tale, it shows how American men have lost the "wild man" within them. Essentially, the wild man is caged up, and the key is hidden under his mother's pillow—"the place where the mother stores all her expectations for you," wrote Bly in the book.

Randy began the quest to "regain the wild man" within, but in the process of that, and writing the Morning Pages, which encouraged looking critically at oneself and one's life, he was taken back to many painful memories: the lack of physical affection in the Morgenson household during his childhood; the final hike with his father, when Randy had failed to tell him he loved him and never truly thanked him for the gift of the mountains; and his mother on her deathbed, with the key under her pillow the entire time.

Two psychologists completed this all-out self-help effort. But so consumed was Randy by the process, he failed to provide any sympathy for Judi on the day her mother died in December 1995. While Judi had been on the phone checking in with a nurse at the hospital, her mother passed away. Judi told Randy what had just happened and that she had to get to the airport, but he didn't offer to come along, nor did he offer her a ride—although he did apologize profusely once he realized how distressed she was. "That shows you how wrapped up Randy was in his own little world at the time," says Judi. "That should have sealed it, but I couldn't untangle myself from everything that was happening at the time, so in a way my mother's death postponed what had to be done just a little longer."

In hindsight, Randy would berate himself for his callous behavior, but for Judi the damage had been done. She tried to hold on to the good memories, the special dinner and bottle of wine he always had waiting for her when she'd hike in to visit him in the backcountry, the evening strolls through the meadows, holding hands in the beauty of the mountains. And there were funny times, such as the night they were sleeping out under the stars and she felt something brush across her face. Randy had touched her and whispered, "Don't move." She didn't, and a giant porcupine ambled away. They sat up laughing as the porcupine rustled off into the darkness.

Now, for Randy, it was all darkness, and worse. He'd brought it all upon himself, and upon Judi. He wrote Scofield on February 10 in response to an invitation to join him at a photo workshop:

> I'm afraid I can't commit at this time. My life is still in upheaval/turmoil (and probably my shutter is rusted shut). Nothing seems predictable, except pain. You've been here. I now understand. One thing that comes out of this is understanding of others. My only hope is that I truly learn something. "We all make it through our garbage, if we make it" said a character in a recent novel. Maybe I'll see you this Spring again, or summer.
>
> —Peace and love, Randy

By spring, Randy had stayed in Sedona for the entire winter and he'd struck upon a plan to try and make it right with Judi. He told her that he had decided to quit the Park Service. What was he working toward anyway? There was no pension or retirement as a seasonal ranger. He'd get back into photography, maybe take on a ranger position in the Southwest. They had some money to fall back on, so why not work on "us." They would kick it off with a springtime sojourn in the desert as they had planned to do four years before. They'd go to the Grand Canyon and hike down a little-traveled trail and soak in the solitude—together. A new beginning.

They arrived at the canyon rim in a snowstorm, and Judi, who was suffering disc problems in her back, didn't feel comfortable hiking down the steep, icy trail. They abandoned those plans and drove toward

Utah, visiting Navajo National Monument, and then on to Monument Valley. They ended up at a tributary of the San Juan River and did an impromptu three-day hike into Grand Gulch. Everything clicked, and they got along wonderfully, wandering slowly through the canyons, Randy taking photos of the amazing cliff dwellings and artifacts that must have surfaced as a result of recent storms. Randy was especially "attentive and sweet," says Judi, who let down her guard completely by the third day and enjoyed the wildness of it all. "It was magical."

Perhaps it was being back in the wilds, or perhaps it was a pang of melancholy at the thought of not returning to his beloved Sierra, that caused Randy to sabotage the romance. As they drove toward Mesa Verde, he backed out on his pre-trip promise to stay with Judi for the summer. With Judi sitting next to him—in love all over again—he mused that he wasn't sure what he should do about their relationship. Judi had lost count of the times he'd broken her heart in the past three years. She realized, finally with clarity, that Randy was never going to make the decision to leave. Before, Judi had been, in her words, "too busy burying my relatives" to deal with his yo-yoing. Now, in the blink of an eye, she knew what had to be done.

She asked Randy to turn around and take her home. In the following weeks, as Randy prepared his gear for his twenty-eighth season at Sequoia and Kings Canyon, Judi prepared their divorce.

He wrote Scofield:

> I've been a mess this winter; the most painful time I can remember. I've spent time talking with two different psychologists, and read more psycho-babble books than I can recall. It's not been particularly fun but it's loosened up some ancient painful stuff that I now have to look at and do something with, so it doesn't keep eating me up. They call it growth. Time to grow up. Meanwhile, back into the mountains.

> —Peace and cheers, Randy

Just before Randy left for his last season, Judi told him she was pretty upset but glad he'd come to terms with the divorce. In return, Randy told Judi he didn't know if he was doing the right thing by going back into the mountains. "He tried to suck me in again," says Judi. "I told him I loved him, but he needed to go, and this time don't come back because I've walked through the door, and there's no turning back for me now."

Randy left Sedona in late May after giving Judi the parting gift of the book *I Heard the Owl Call My Name*. In his backpack was a folder with their divorce papers fully in order. He told Judi he wanted to clear his head in the mountains before he signed them. She didn't push the issue.

Before ranger training began, Randy stopped in Bishop, seeking to take counsel with his old boss and, hopefully, still friend, Alden Nash. When nobody answered Nash's door, Randy walked into the backyard through the side gate he'd used for years. Nash's home was as familiar as his own; he'd stored gear in its garage for two decades and nursed a pinched spinal nerve on its living room floor for weeks one season. A drive through Bishop wasn't complete without a visit to the Nash residence.

Nash was over in a corner of the yard, weeding a flower bed. In the past, he would have stopped whatever he was doing if one of his ranger alumni came by for a visit, but he wasn't happy with Randy, who had, in Nash's words, "been living two different lives, and I wasn't sure if I could take him seriously." Uncharacteristically, Nash barely paused in weeding as Randy spoke.

As if he'd rehearsed this moment a hundred times, Randy apologized to Nash for not being straight with him, admitted that he'd felt ashamed about the affair, that he'd hurt people he cared about, and that now he was probably getting what he deserved. Mostly, Randy was seeking advice from the man who had always been there for him. Randy nearly broke down when he said, "Alden, when I wake up in the morning and look in the mirror, I don't like who I'm seeing." At that Nash gave Randy his full attention and a deep stare.

"What can I do about that, Alden?" he asked.

Nash looked off toward the towering Sierra to the west and then the White Mountains to the east while he thought through his response. "Well," he finally said, "first of all, you have to make things right with Judi. That's a good place to start."

AT TRAINING, Randy was unusually subdued and made the rounds to many of his friends and delivered well-thought-out apologies for his behavior, not only recently but also over the years.

And then he called Judi and said he had thought about everything on the long drive from Arizona. "It doesn't feel right," he told her, "going into the mountains without having you to come home to." He asked her if she'd drop everything, throw together a pack, and spend the season with him. "I'm stationed in a beautiful spot," he said. "Bench Lake, high, windy, not too many bugs."

Judi remembered a time when Randy could make a mosquito sound romantic, but her answer was still no. She loved him dearly, but she wasn't a pushover. He continued with "Well, how about for a visit. I can meet you at the trailhead, and . . ." Her mind was made up. The answer was no.

Randy waited for the right moment to approach Lyness. He wanted to know if there was any chance of a future together, or if he'd messed that up too. By now Lyness was devoted to her boyfriend, which was news to Randy. Her answer, too, was clearly no.

For the rest of training, Randy gravitated toward the newer backcountry rangers—Rick Sanger, Dave Gordon, and Nina Weisman, whom Randy had encouraged to follow her dream to become a backcountry ranger throughout her career at Sequoia and Kings Canyon. At one point during training, she thanked Randy for his guidance over the years. He was surprised and asked her, "What did I do?" She reminded him of the time when he had talked her out of being lost—but mainly, she said, she had been inspired by his example. Another fairly new ranger named Erika Jostad gave Randy a talisman she'd carved from some downed wood she'd found the season before in the backcountry. She had sensed he was in a funk and thought he would appreciate the gesture.

"I think Randy thought he had alienated most of the rest of us," says George Durkee. "But the truth was, we just wanted everything to get back to normal."

ON JULY 1, 1996, Randy assisted helitac (flight crew member) Carrie Vernon as she unloaded his provisions at the Bench Lake ranger station site. Cardboard boxes, a weathered backpack, duffel bags, a ski-pole hiking stick, and a crate of citrus fruit—the usual.

The helicopter lifted off, and after its wash subsided, Randy stooped to pick up a large flake of black obsidian from the loose soil. He showed the arrowhead-in-progress to Vernon and remarked on what a spiritual place they were in.

Usually, a helitac left with the pilot after unloading gear, but on this insertion, Vernon stayed behind while the helicopter flew back to Cedar Grove to pick up Cindy Purcell, Kings Canyon's new Sierra Crest subdistrict ranger. Purcell was to be Randy's supervisor for the season and intended to help him set up camp.

Vernon immediately began carrying boxes to the 12-by-15 plywood platform that would become the floor of Randy's home once he put together the cabin-style tent that had been stored for the winter in a 50-gallon steel drum. As Vernon schlepped gear, Randy dragged from a nearby stand of evergreens a picnic table, where it had been turned upside-down for the winter. He oriented it for the optimum view of Arrow Peak, then ushered Vernon over and offered her a seat.

"No rush on that," said Randy, in regard to his gear. "I've got all season to get organized."

"That was what struck me as unusual," Vernon says. "Normally, Randy was running around like a kid the second he got off the helicopter, hugging trees, scouting out the water supply, and checking how everything fared over the winter. You would almost have to stop him from climbing some nearby peak right off the bat instead of helping to unload the helicopter. But he just wanted to talk. I'm not saying he didn't normally talk—if you got him on the right subject, he'd talk your ear off—but at the beginning of the season he would normally . . . I don't know . . . kind of give you the impression that he'd rather you get out of there and leave him alone."

Vernon spent almost three hours chatting with Randy on that picnic table in what she describes as "one of the most pristine spots in the park." The two talked about all sorts of things—the backcountry, the Paiutes, the snow on Pinchot Pass, and about how being a ranger was tough on relationships. "He asked me for advice," remembers Vernon, "which was kind of out of left field. He was trying to figure out how to strike a balance between being a good ranger and being a good husband. Something like 'How can I be in here and at the same time be a good husband out there?'"

Eventually, they decided that there was no magic recipe, that you had to follow your heart and do the best you could. When you're in the backcountry, be a good ranger. When you're at home, make up for lost time and be a good husband.

Randy knew he was a good ranger. The impression Vernon got was that he was also going to make a damn good effort to be a better husband as soon as he finished the season. He even hinted that this might be his last season, if that's what it took. "Might be time to try something new," he told her. Vernon couldn't imagine anything that would suit Randy better than this job. But she knew, and so did Randy, that backcountry rangering took a toll on a person's body.

Seasonal rangers either quit or they die. There is no retirement option, at least in the traditional sense in which your employer supports you through the golden years. Randy had told more than one of his colleagues, including Vernon, "If I retire, it might mean a party. Maybe I'll get a plaque. But there sure won't be a pension."

When the helicopter returned, Subdistrict Ranger Cindy Purcell exited with a backpack and the few remaining boxes that constituted the Bench Lake ranger station summer rations. Purcell was a permanent ranger—full time, year-round. But if there was one thing Randy had learned, it was how temporary the permanents are. He had in fact trained many of his bosses, at least in regard to "the resource." Randy hated the term "resource." To him the resource was his home, which was the reason why Purcell had chosen Randy to show her around.

"When I met Randy during training, you could tell he had a kindred sense of ownership to the High Sierra," says Purcell. "I wanted to spend a few days with him so I could better understand the backcountry issues and the rangers I'd be supervising."

Randy told her right off the bat that the best way to understand the rangers' way of thinking was to spend time in the mountains. "We could sit here at this picnic table all day and talk about it," he said, "but you'd be better off exploring some on your own." As he'd once written many years before, "Find it yourselves, and it will be all the sweeter."

As Randy spent three days organizing his station, he started Purcell off on different trails branching out into the wilderness. One day, he sent her north to Mather Pass; the next, east to the top of Taboose Pass. From the station, Taboose was a casual ascent on worn trails across meadows, over creeks, through wooded glades, and near various archaeological sites that Randy had uncharacteristically mapped out for Purcell to investigate.

The view from Taboose Pass, looking east across Owens Valley to the Inyo Mountains and Death Valley beyond, is one of the most dramatic in the Sierra. Two worlds converge there at the crest: cool, green, blossoming high-country meadows, still spring in July, and, 8,000 feet below, the rust, white, and swirling dry heat of the desert with 100-degree-plus temperatures blurring the horizon. Alone on the edge of these two worlds, Purcell felt small in the vastness of it all. It was exactly the feeling Randy wanted the experience to evoke. He hadn't told her what to expect, but instead advised her to walk slowly "and look around once in a while."

On the Fourth of July, 1996, Randy accompanied Purcell south from his station on the John Muir Trail. As they neared the snowy crux of Pinchot Pass, he acquainted her with his favorite mountain flower, sky pilot, which, he wrote in his patrol log, was "perfuming the air." Another discovery near this formidable pass was a trailside pipit nest holding four tiny brown eggs. Before the early 1980s, pipits weren't known to nest in the Sierra. Then Randy discovered them near Tyndall Creek and Wright Lakes, adding them to the list of resident wildlife protected in the parks. During Purcell's short stay at Bench Lake, Randy bombarded her with so much casually delivered natural history, she felt as if she'd taken a crash course in High Sierra, taught by John Muir himself. Purcell didn't realize that Randy had spent more time in the Sierra than even Muir had.

At the Pinchot Pass summit, Randy parted ways with Purcell, who was continuing south 15 miles to meet up with the next backcountry ranger on the JMT, Rick Sanger. As Purcell walked the trail, she felt exhilarated by her new job. She felt fortunate to be commuting to work on a high-mountain footpath, and she understood why Randy had been coming back for so many years. It struck her that she'd "never met a man who was so genuinely enamored by the mountains and so at ease in such a wild place." This was something she aspired to.

Sixteen days later, on July 20, Randy radioed over to LeConte Canyon and spoke with Durkee and his wife, Paige Meier—a conversation they originally interpreted as "Randy just wanting somebody to talk to." The short conversation ended when Randy said abruptly, "I won't be bothering you two anymore."

The following morning, according to the note he left at his station, he went on patrol.

CHAPTER TWELVE

THE RANGE OF DARKNESS

It must have been a vexing disappointment. Where the devil could Starr have gone? Maybe the others had been right—it was proving to be like hunting for a needle in a haystack.

—*William Alsup,* Missing in the Minarets: The Search for Walter A. Starr, Jr.

One hiker limped and wobbled up the valley in late afternoon, passing like a ghost through the lodgepoles. Did I really see someone, or imagine it? Dreaming? The mountains, and their companions the forest and meadows and evolving creek just stand here. They're not telling.

—*Randy Morgenson, McClure Meadow, 1973*

FOR THE OVERHEAD PLANNING TEAM, July 31, the seventh day of the search, was pivotal. The day before, there had been ninety-eight personnel. This day that number was reduced to ninety: from forty-eight ground crews to forty-two; from four helicopters to three; from eight dog teams to six. Up until now, the numbers had risen daily. The decrease, however slight, spoke volumes. "You could sense that hope and optimism began to decline," says Scott Wanek, who served as both the safety officer and operations section chief throughout the SAR. "The reality that the search had to be scaled back was hanging over everybody."

There were those who held on to the vision that Randy was out of the mountains, running, probably in a bad way, but alive. But for many of the backcountry rangers, and for Chief Ranger Debbie Bird, there was little doubt that Randy was somewhere in the park. She knew him well enough to be absolutely certain that "he would not do this to his fellow rangers. I never believed he was the type who would have willingly exposed his closest friends and colleagues to something like this."

Randy Coffman, who as incident commander couldn't let his personal opinion cloud his judgment, was steadfast in his belief that Randy had not taken his own life. That didn't mean Coffman hadn't acted upon the suggestions of those who did believe in the suicide theory, such as highlighting the bases of cliffs as places to search. Gallows humor among some volunteers dubbed this the "swan dive" theory. Nobody close to Randy found any humor in the joke.

Ultimately, the burden of scaling back or ending the search rested on Coffman, who had to weigh the risks of the search itself against the probability of locating Randy alive.

Already, search teams had been chased off summits by lightning. One helicopter negotiating a tight landing zone had clipped a tree, while another had had a "hard landing" due to high winds and had to be withdrawn from the search for repairs. An overzealous search dog had bitten a ranger; a loose rock had crushed another ranger's hand; and there had been various close calls: rock slides had come dangerously close to ground crews, and a searcher had punched through an air pocket up to his waist while crossing a snowfield. Volunteer searchers from sea level had been evacuated for altitude sickness. One trail-crew supervisor had had such a frightening flight that he was uncertain whether he'd ever get back on a helicopter.

Fortunately, there had been no serious injury to the human searchers. The search dogs, on the other hand, hadn't fared so well. Most lasted only a day before being incapacitated by paw lacerations. Seeker had almost drowned.

Even with the massive amount of air and ground activity in virtually all of the search segments, there was still the chance that Randy was seriously injured somewhere and unable to signal to the constant flux of helicopters because of a radio problem. This kept hope alive in the public forum, but behind closed doors the inevitable was upon Coffman and Bird. If it had been a park visitor, this—the seventh day of the search—would likely have been the last. But Randy's survival skills made it conceivable that he could still be alive. Bird, as the ultimate supervisor of all the rangers, also wanted to make sure Randy's friends and coworkers knew they had done everything possible before calling off the search.

BOB KENAN AND CHARLIE SHELZ woke up on July 31 in the Window Peak Lake drainage. They'd camped in the open on a sandy flat among giant slabs of granite, and after a rushed breakfast they continued down the rugged talus fields and glacial moraines dominating the landscape. They were still focusing on the west side of the creek that drained into Window Peak Lake. Kenan knew this area well; it had been one of his favorite cross-country loops when he was stationed at Bench Lake in the mid-1980s. There are two standard routes from the lower basin to the inflow of Window Peak Lake, both of which are accessed about a quarter mile above the lake.

They form a sort of Y intersection at the most logical crossing point of the stream—a shallow, flat-water area above a small waterfall, 50 yards downstream from the lake Seeker had fallen into the day before. From there, you can either access a loose-gravel gully on the east side of the creek that leads down to the lake or follow the creek itself, which enters a granite chasm and drops in elevation fairly rapidly over a series of small waterfalls. During the search, the chasm was full of late-season snow and ice, but normally it would have been the type of gnarly route choked with willows and brush that Kenan couldn't resist.

Kenan had, in the past, called this lower section of the chasm the Gorge of Death, in reference to one of his more intimate encounters with the wildlife of Kings Canyon around 1987. "I was thrashing through brush and willows over my head, my mind focused on the lake and the big fish I intended to catch for dinner," he says. "It definitely wasn't an inviting route, at least for a human, but it was a likely spot for a deer to get a drink of water and eat the grass that grew thick along the creek.

"I'm certain that's what the mountain lions mistook me for. I pushed through a final section of willows and came face to face with two lions ready to ambush whatever it was making all the ruckus. That gives you an idea of how remote that area is. Anywhere near a real trail, a mountain lion would get spooked by the kind of noise I was making. It was a mother, with a juvenile in training. They were about 6 feet away and ready to pounce. But the mother leaped away and then stopped and looked back at Junior, who was trying to figure out whether or not I was food. It just sat there staring at me—an amazing, beautiful animal.

"I didn't know whether to make a bunch of noise that might have alerted the mom that I was attacking the little one, but luckily I didn't have to. Junior reacted to Mom and sprang away, and I was left there with a shot of adrenaline straight to my heart. From then on I always made it a point to talk out loud and sound human, not like a deer, when I took that route—but usually, I took the easier gully route down the east side of the creek, bypassing the gorge altogether."

Kenan and Shelz weren't able to cross from the west side of the creek they'd been searching to the east side where the easier route was because of the snow clogging the ravine, the result of avalanches and heavy snowfall. The snow also made the gorge itself an impossible passageway down to the lake. They took a route along the west side of the creek Kenan had never taken because it was a long, steep cluster of broken and loose talus—a difficult route that proved to be even slower because they'd been carefully implementing a tedious, line-of-sight search technique for rough, erratic terrain: walking a few steps, stopping, and looking back at the route they'd taken and then in every direction. This allowed them to cover all vantage points that might reveal a clue hidden in a crag or camouflaged by a shadow. A clue that might have been invisible just a few steps back.

So sure were Kenan and Shelz of their search methodology, they wrote in their debriefs that if Randy was mobile in this area, they were 100 percent certain they would have found him. If he was "immobile but visible," they gave a score of 50 percent. Even if Randy were immobile and not conscious, they rated their "probability of detection" at 20 percent. Under "problems encountered with communication," they wrote "none." If anything, there was "too much [radio] traffic." Despite the remoteness, the basin's orientation to one of the parks' radio repeaters on Mount Gould created, in effect, a radio signal channel. Under Suggestions, they wrote: "Thoroughly covered area; would not go back into this area."

Rick Sanger, meanwhile, had been joined at Window Peak Lake by CARDA volunteer Eloise Anderson and her black Labrador, Twist— the dog team that had been called to pick up where Linda Lowry and Seeker had left off. In 2003, Anderson and Twist would gain national attention as one of three dog teams called upon during the Laci Peterson investigation; Twist's skills as a cadaver-trained search dog would provide the prosecution with valuable scent-based evidence. At the Morgenson SAR, Twist had yet to become cadaver-certified.

Anderson's assignment with Twist was to continue down the Window Peak drainage until they hit the John Muir Trail. With Sanger, the dog team cleared a small area to the north of the lake, then headed down-canyon to the trail. The wind, like the day before, was blowing up the canyon. Twist did not express interest and did not alert once.

IT HAD BEEN TWELVE DAYS since Randy's last contact, and the search effort was further scaled back to seventy-five personnel (including thirty-two ground crews), four dog teams, and three helicopters.

On this, the eighth day of the search, Durkee noticed that Graban had a "thousand-mile stare" and requested to be her partner. "I wanted to be there for her—you could tell she was pretty fried," Durkee says. In retrospect, he admits, "Maybe I needed Sandy's calm nature to hold on to."

The two were dropped off at the lowest Dumbbell Lake by military helicopter, to search the area between Dumbbell Lakes to the confluence of Cartridge Creek, a steep drainage alive with willows, brush, and loose and slippery rock that had been searched already. Graban described it as "heavy bushwhacking."

Looking into the gorge, with the helicopter taking off behind them, Durkee was overcome by the spirit of Joseph Conrad. "We're descending into the Heart of Darkness here," he radioed in to the incident command post. "Will report at the bottom." Durkee had bantered similarly with Randy many times in the past. Those were the good old days of deadpan ribbing and carefree rangering despite the seriousness of their jobs. Like the time Durkee flashed Randy the pink V-neck "Jane Fonda Workout" T-shirt he'd put on under his ranger uniform while on standby when the actress was reported missing in the park's backcountry during the early 1980s.

Or when each would try to better the other's "traffic reports" of backpackers on the "John Muir Freeway" or the "John Manure Trail": "Looks like there's some heavy congestion around the JMT/Kearsarge interchange, with hikers backed up all the way to the Bullfrog overlook. We suggest you take the Charlotte bypass to avoid the mess. This is 115 aboard LiveCopter 52. Jump and jive with 1-1-5. Back to you in the studio."

But nothing compared to Randy's legendary "Conversation with a Coot." A coot, all Sierra rangers know, is an inquisitive duck that makes the high mountain lakes its home for a brief period each summer. Randy had been "interviewed" by a coot, or so he reported. The curious duck had asked the same stereotypical questions Randy had fielded from backpackers over the years.

"So, how'd you get this job?"

"Is it lonely out here all alone?"

"How do you get your food?"

"Do you have to stay out here all summer?"

To which Randy replied, "Duck, I *get* to stay out here all summer."

"What do you do the rest of the year that lets you take your summers off?"

To which he showed the duck his pack full of backpacker garbage he'd collected that afternoon, "Well, this isn't exactly a summer off."

Randy had actually been interviewed by a few different species of duck, and even dined with a chipmunk one year—but it seemed the coot's line of questioning rang most nostalgic to Durkee, who, at this point in the search, couldn't stop shaking his head at the memories. Was it possible that these recollections were all that remained of his friend?

With a wink at Graban, and their final assignment of Cartridge Creek beckoning, Durkee signed off with the ICP dispatcher, "The horror! The horror!"

Then the two veteran rangers trudged forward, Durkee's monologue echoing off the granite walls: "Going up that river was like going back to the earliest beginnings . . . an empty stream, a great silence, an impenetrable forest . . ."

Other than a couple more inconclusive tracks, one of which had belonged to a hiker with a long stride—too long for the 5-foot-8-inch Randy—nothing came of their bushwhack through the "Gorge of Darkness." Randy had vanished. The mountains had swallowed him up.

The loose ends were being tied off. Virtually all the segments had been covered by air and at least one ground team with a dog. Now as many areas as possible were being checked for a second, sometimes a third, time. Near Durkee and Graban, a dog team was in vertigo country—dizzying, you-fall-you-die terrain south of Amphitheater Lake. The theory was that in working a 12,000-foot ridge with a search dog, updrafts would reveal if Randy was down either side of the spine—a classic strategy used to expose a dog to a large area of scent. However, the mountains weren't cooperating. Zero updrafts left the canine focused on the narrow ridgeline itself, so the searchers did their best to down-climb a few sections that seemed possible places for a slip. Progress was always stopped by cliffs. "The only way to more thoroughly search this area," the team members wrote on their debrief, would be "with a rope."

SEKI backcountry rangers Dario Malengo—who was back on the search with a cast on his hand—Rob Pilewski, and Rick Sanger, working with four dog teams over the previous couple of days, had covered what amounted to hundreds of miles of terrain. Again, without a single clue.

"I don't want it to sound weird," says Sanger, "but it was like we were looking for a ghost. Randy was super low-impact. He didn't leave a trace when he camped, but we covered all the logical routes from the Bench Lake station and there was nothing. It was frustrating because my gut told me he was hurting out there and needed our help. I couldn't ignore that sense on the eighth day of the SAR. I let myself accept that Randy had probably died alone out there. I just hoped that he didn't suffer, or worse, that he was still suffering. I prayed that it went quick, whatever happened."

That afternoon, Lyness prepared to vacate the Charlotte Lake ranger station. It had been planned well in advance that she would leave the mountains early in the season due to her frontcountry job. Randy, who would have taken over the Charlotte Lake station, had cached some food and gear there—haunting reminders. Now with Randy missing, nobody knew who would fill the position.

Word had gotten around: if no concrete clues were found that day, it was likely the search would be called off. "No apparent progress on search," wrote Lyness in her logbook. "Got a call this afternoon that I'm to be flown out tomorrow and my stomach's in a knot. Afraid the debriefing will be worse than what's gone on up to this point. Crying hysterically in front of a group of your peers is not at the top of my list."

At day's end, Durkee and Graban were flown to the incident command post at Cedar Grove. They ambled halfheartedly to the fire station's planning room, where the overhead search team had been based. Inside, two picnic tables were pushed together. At one end, Coffman sat with a stack of papers, calm and composed but uncharacteristically subdued. A dry-erase board off to the side featured a list of notes, with two stick figures labeled "Durkee and Graban" climbing down a cartoonish cliff along the edge of the board. The taller stick figure was saying, "The horror! The horror!"

"How appropriate," thought Durkee as he sat down at the opposite end of the table from Coffman.

The seats between them filled with other members of the ICP's overhead team, including Dave Ashe and Scott Wanek. Debbie Bird was there as well.

Coffman began the meeting by thanking everybody for their hard work and then got down to business, going over the clues they'd followed up on.

A bag of trail mix was found in an abandoned campsite (the mix didn't match any food Randy had at his station). A First Need water filter was discovered near Bench Lake, where the smell of "something dead" had been reported by a search volunteer (Randy's water filter was still at his station, and a cadaver-trained dog did not alert in the area). A man was reported crouching in a snow cave on a high slope (the man proved to be a shadow). Something had glimmered from a talus slope on a peak, perhaps a signal mirror (it was a discarded water bottle). Numerous tracks were noted (scent-specific tracking dogs determined that they were not Randy's). In all, around two dozen clues were documented; none had led to Randy.

Coffman was well aware of the hazards encountered by the search teams—steep terrain, cliffs, swift water, bedrock "slick as snot"—and he went over a few of those scenarios as well. Durkee saw the writing on the wall: "He was breaking it down for us, showing us a little of his thought process, so we'd understand why he had to call it off."

Before this meeting, Coffman had privately talked it over with various rangers who were closest to Randy. Was there anything they should or could do that they hadn't already tried? "I was extremely thankful for that gesture from Coffman," says Durkee. "Asking me my opinion was probably one of the most gracious gestures of respect I ever got from a high-level administrator."

Coffman queried this group as well, challenging them to think. As could be expected, he—all of them—were pretty fried. "The overall mood at that meeting was grim," says Durkee, "and when nobody came up with any new ideas, Coffman said he'd gotten something from a psychic, who suggested where Randy might be."

When Coffman started to read the psychic's letter, Durkee began laughing, semihysterically. "Perfect," he thought. "Randy would have loved this moment in his search." But when he realized nobody else had joined in, Durkee abruptly stopped. "Oops," he thought. "They're serious about this—Coffman is serious about this."

With a sigh, Coffman looked everybody in the eye. "And to his credit," says Durkee, "he almost sheepishly suggested that we were out of options." Coffman then focused on Durkee and asked him directly, one last time, if he thought it was time to stop the major search effort. With the spotlight on him, Durkee leaned heavily on his elbows, his chin resting on clenched fists. After a long silence, tears welled up in his eyes. Durkee nodded his head, and then he began to cry.

"A long, strange trip," says Durkee, remembering that moment. "Total bungathon—but there was no other choice."

Operations Section Chief Scott Wanek had, for the past couple of days, hedged his vote when Coffman asked him what he thought about stopping the search. He admitted later what everyone assumed was a possibility: "that we'd end the search and then find Randy sometime later with a note documenting his last days, dying after we'd called off the search from some injury he'd been suffering. It's the 'What if?' factor that makes calling off a search so emotional. I hate to admit it, but the decision would have been easier if it had been someone from the general public."

Bird echoes this sentiment: "The truth is that had it been a visitor, we would have scaled back the search a full day or two before we did."

By the end of the meeting, however, "everyone agreed that to continue to search would be essentially fruitless," says Bird. "In reality, we sustained the search because I did not want any of my employees thinking that we did not do everything we could have done." Given the rangers' feedback, Coffman and Bird jointly made the decision: the Morgenson SAR would end the following day.

On August 2, thirteen days after Randy's last contact, radios crackled to life across the Sierra wilderness. Nina Weisman was on the porch of her Bearpaw Meadow ranger station sipping hot tea; Lo Lyness was at Charlotte Lake; Rick Sanger was at Bench Lake awaiting his next assignment; Graban and Durkee were at Cedar Grove, having stayed the night in the employee-housing cabins there. The grave voice on the radio was that of Sequoia and Kings Canyon superintendent Mike Tollefson: "The search for Randy Morgenson has occupied our hearts and minds for many days now," he said.

"Randy last checked in by radio on Saturday, July 20. We know that he talked to two hikers that day near Mather Pass and that he returned to his station at Bench Lake that night and made an entry in his log. When there was no radio contact for the next forty-eight hours, a ranger went to check on him.

"Randy had left a note at his station on the twenty-first stating that he had gone for a three-day patrol. A hasty search was initiated on the twenty-fourth, using a helicopter and several staff, and a full search operation began the next morning.

"The effort put forth by people from all over these parks, as well as other parks and agencies, has been immense. They searched a rugged 80-square-mile area with almost one hundred people, five helicopters, and eight dog teams. But there has not been a single clue.

"Based on this intensive search, and the absence of any leads, we must begin to scale back on direct search tactics. Nonpark resources are being released to their stations. Park personnel will continue searching on Friday and Saturday. If there are still no results, we will scale back further.

"A ranger will be stationed at Bench Lake for the rest of the summer, interviewing all hikers and continuing to search. At trailheads we will continue to ask backcountry visitors to be watchful for clues during their trips.

"To know that Randy is missing is difficult; to have no resolution to the search is more so. For those involved in the search, there will be a critical incident stress debriefing over the weekend: contact Debbie Bird or Randy Coffman. Also, the services of the Employee Assistance Program are available to all.

"Deep thanks are due to those who participated in the search, and our thoughts have been with them throughout.

"We will continue to seek the answer, as we will continue to keep Randy in our hearts.

"Thank you for being special."

Then the radios—for the first time in ten days—fell silent. The official search was over.

AFTER ATTENDING the group peer counseling session or individual counseling, the backcountry rangers returned to their duty stations in the high country. Critical incident stress management professionals call what these rangers had just gone through a "mission incomplete." In essence, not finding Randy, a fellow ranger, was one of the most traumatic experiences to process and often leads to classic post-traumatic stress.

Individuals who have experienced a "mission incomplete" are, in a perfect world, routinely followed up on after their initial debriefing. For the backcountry rangers, there would be no follow-up.

Rick Sanger was flown back to the Rae Lakes ranger station. As the helicopter lifted off, he lay down on the ground and waited for the rotor sounds to fade away. He hoped that he wouldn't hear another helicopter for a long time. Eleven days had passed since he'd left his post to check in on Randy the evening of July 23. He remembered the hike, his lack of concern, and the happy anticipation of finding Randy with a broken radio—of boiling tea and catching up.

Now he wasn't sure if he was exhausted, depressed, or in shock. Feigning motivation, he tidied up the station. He wiped the dust and mouse droppings off his tiny desk; glued the sole of his boot, which had started to separate; and recharged his HAM radio battery with solar panels. Though physically tired from the search, he had to walk somewhere. He wasn't sure where.

Once he got going, an unnamed lake the rangers call Ranger Treat Lake seemed a fitting place to dissect the emotions Sanger was feeling. The hike, he reasoned, would do him good. He stayed near the lake's shore until nightfall, waiting for the soothing alpenglow on the granite, hoping the Range of Light would work its magic and heal his wounds. It didn't.

He walked home in the dark.

That night, Sanger pounded the keyboard on his laptop—likely the only computer in the entire backcountry—inserting notes from the pad of paper he scribbled on obsessively. Eventually he caught up to the present day: "Cruise up to Ranger Lake. That place is a fucking temple, unbelievably beautiful. The shots of light passing on the lake's surface, the edge of the lake suspended in front of Fin Dome. The placement of the trees and their shapes.

"The hardest part: the best I could give was not good enough. Should I have gone longer hours? Hiked farther? I envision it's going to be spooky back here. I had nightmares of discovering him, and of him staggering into camp and falling on my tent."

The glow of the laptop's monitor competed with the hissing Coleman lantern hanging from the ceiling to cast an eerie light. "Staring down at carcass of moth and mouse turds on floor," Sanger typed, his mind spinning back to the critical incident stress debriefing. "Finding it tough to look anyone in the eye. 'Be careful of alcohol' [one of the peer counselors had warned the rangers].

"I'll be careful to drink every can I come across."

Sanger closed his laptop, hung his headlamp near the foam mattress that served as his bed, turned off the lantern, curled up in his sleeping bag, and retraced every route he'd taken in the search. He blasted himself for every moment he had not been on his feet looking for Randy.

The next day was an "administrative leave" day for the rangers who had taken part in the search. "Slept late, read Ed Abbey's *Fool's Paradise*, made pancakes, and tried to soak in the events of the past week," wrote Sanger. "Sat out in the shimmering sun with the onions that are now blooming around the spring.

"Reading and staring at Painted Lady and the Crest. This is an irrec-oncilably beautiful place. Abbey found his niche with the park service and I chuckle at a deep level of understanding at his descriptions.

"How lucky am I to be here?

"Why can't I hold myself in higher regard and admiration for where my body and spirit have led me!"

Meanwhile, just over Glen Pass at Charlotte Lake, Lo Lyness was equally somber:

Saturday, August 3, 1996, was "a painful day," wrote Lyness. "Critical Incident Stress Debriefing. It was okay. I just feel like a truck has run over me most of the time. . . . Randy would never believe how many people have been touched by this, how many people care and want to help, how many people have and are hurting.

"Flew home . . . told to take the rest of the day and the next off. So I did. Read, wrote, cried. Sat by the lake and looked at the mountains. Tried to understand <u>something</u>. But it's not understandable."

The following day, Lyness slept in late. She picked up a book she'd brought into the backcountry, *Everett Ruess: A Vagabond for Beauty*, by W. L. Rusho. Describing it in her logbook, she wrote that Ruess "disappeared in the Southwest at a far younger age than RM; never to be found."

She and Randy had talked about Everett Ruess in the past, but she can't recall exactly why she was reading it that season. She calls it a "crazy coincidence" that Randy had been reading the same book just before he disappeared.

As her season continued, she tried her best to soak in the back-country: "fields of purple and white," she wrote, "more gentians than I've ever seen in one place . . . it's magic." On August 6, five separate hikers asked Lyness about Randy. "Wonder how long until it doesn't feel like a knife in my stomach whenever someone asks," she wrote.

On August 12, Lyness headed out on a long patrol to the Upper Kern via Forester Pass, where she cut off the trail, stashed her pack, and started to scramble up Cal Tech Peak. That evening she camped at a favorite ranger patrol site, the little frog pond en route to Lake South

America—likely, the same hideaway camp Randy had taken his father to on his last trip into the high country. In the night, Lyness was awakened by "a nightmare about Randy" and couldn't fall back to sleep. "A blessing in disguise," she wrote, "as I saw lots of meteors—it being the Perseid time."

On Lyness's final day in the backcountry, it was a "gorgeous, cool, Fall-ish day. Hard to leave now. But—off to join the rat race." That was the last entry Lyness made in a station logbook. She never returned to the high country in the uniform of a backcountry ranger.

George Durkee returned to LeConte Canyon, where his wife, Paige Meier, had held down the fort, acting as a radio relay during the search and den mother for searchers passing through.

Two weeks later, Durkee was still in a fog. "Objectively," he says, "I knew Randy was dead, but emotionally, I hadn't processed it. I'd wake at 0300 thinking, 'Hmmm, nobody ever checked *there*.'"

Durkee found he wasn't the only one still "processing." Backcountry ranger Dave Gordon had a "flying dream" in which he was searching for Randy, "and just as he was soaring over a lake," recounts Durkee, "a drop of water from a tree hit him on the forehead and woke him up. He took it as a sign of where Randy was." The lake in Gordon's dream was north of Window Peak. The story got around to backcountry rangers Rob Pilewski and Larry Stowell—who searched the lake to no avail. In this manner, the rangers continued to pick and poke their way around the mountains looking for Randy.

Then one morning, around three weeks after the search, Durkee and Meier were washing dishes in the clearing in front of the cabin. Durkee was staring north up the canyon when he thought he saw a figure approaching through the trees. He was midconversation, rinsing dishes in a basin, when everything got "hugely weird."

"I had an extremely hard time separating reality from the humanoid figures that were approaching from the trees," he says. "When they came out of the trees and started walking across the clearing past the outhouse, I asked Paige, 'Am I making sense?'"

Meier took one look at her husband and knew something was wrong. Perhaps it was the slight sway in his stance, or maybe it was his eyes tracking around him, intently watching something that wasn't there.

"When I started hearing voices, that's when I picked up the radio," he says. Soon Debbie Bird was informing Durkee, "You're fine; we're sending a helicopter."

Before dark Durkee was out of the mountains at the Bishop Airport, being whisked "embarrassingly" to the hospital, where he was poked and prodded for a couple of hours. The ER doctor listened closely as Durkee told him what he did for a living, about his friend's recent disappearance, and vividly described his hallucinations. By the end of the exam, Durkee had spelled out the emotional stress he'd been subjected to, not only recently but also throughout his career. The doctor came back with "Did you, by chance, eat any wild mushrooms?"

A second doctor with less clinical persuasion and some psychological background pegged the episode as classic cumulative post-traumatic stress.

Over the years, Durkee had seen too many corpses in the backcountry, some of which he had sat with through the night alongside grieving relatives. Randy's disappearance had been a daily dose of anxiety that grew exponentially. Each talus field Durkee had approached might have revealed his friend's body. As the doctor put it, "Your friend's disappearance was the straw that broke the camel's back."

Durkee was most concerned with a recurrence of the hallucination, which, the second doctor explained, was "likely."

Durkee took advantage of his time in the frontcountry to check in on Judi Morgenson. He explained to her his "episode" in the woods around LeConte, recounting how Randy had always said, "That place is full of spirits."

Judi offered her own diagnosis of what had happened. "George," she said matter of factly, "it was a panic attack. I get 'em all the time now."

For Durkee, just knowing he wasn't going crazy was a gigantic relief. He was sent back to his duty station with strong convictions: the antithesis of closure.

"This thing isn't over yet."

NOT LONG AFTER Durkee's episode, Judi Morgenson received a letter at her Sedona home from the Yavaipai County Court in Arizona—postmarked two days after Randy had left his ranger station to go on patrol.

It took just a second to skim the first few lines and find a familiar name. James Randall Morgenson had requested to reconcile the divorce proceedings pending with the court. She recognized his signature.

Judi got light-headed and had to sit down.

Then she gathered her wits and dialed a phone number she'd memorized, the direct line to Special Agent DeLaCruz, whose job—after the search ended—had just begun. Not only did he have to take all of the information from the search and summarize it for his law enforcement investigation report, he also had continued to pursue both the missing-person angle and the possibility that Randy had been the victim of a crime.

The question Judi had for him was simple: "How?" How could Randy have contacted the court and requested to reconcile if he was dead somewhere in the backcountry.

DeLaCruz had no immediate answer. But, he assured Judi, he would. It would just take a little time, which always seemed to be in short supply for his understaffed, underbudgeted criminal investigation team, but DeLaCruz was determined.

With the help of his team, DeLaCruz had crossed both Packer Tom and Doug Mantle off the list of potential "foul play" suspects. Around July 21, Packer Tom had been in New Mexico and Mantle had been climbing Mount Reinstein—far from the Bench Lake area, though still in the parks—with three others. His whereabouts were confirmed by the pack outfit that he'd hired for the trip.

When Mantle learned that he had been a potential suspect, he was at first "stunned" and "dumbstruck." Sure, he'd been pissed off about the citation Randy had given him for his companions' improper food storage. Sure, he'd voiced his anger in the *Sierra Echo* article. But he wasn't a murderer. After finding out that the investigator who checked his whereabouts had made it clear the NPS "had to check everything," Mantle's response was "fair enough."

Later, after completing the Sierra Peak Section list of 247 peaks for the fifth time and becoming the first to do it solo, he wrote a humorous account of the achievement for the *Sierra Echo,* poking fun at himself while, for example, recounting his ascent of Mount King: "But my flashback is of my lunge, and both arms groped about an outcrop, feet kicking freely, about five feet to the left of the correct chimney. I was worried most at that point about being seen."

In the next paragraph, however, Mantle wrote of the challenges he had to overcome while climbing in the Sierra, how "The Range itself tired of hurling obstacles at me, like winter, Rangers Randy (I did not kill him) and Kamenchek (if I had the chance) . . ."

Durkee, among others in the ranger ranks, felt the remark was "tasteless and uncalled for." He wrote the editor of the *Sierra Echo* voicing his displeasure. When he received a letter from Mantle in return, Durkee filed it unread in his "Randy" file, not caring whether it was "an apology or a justification." (The letter was, in fact, an apology stating that Mantle had intended his comments to be facetious.)

After a few painfully long days, Judi learned from DeLaCruz that the Yavaipai court had received a written request from Randy postmarked July 22, one day after he had presumably gone on patrol from Bench Lake. How could Randy have mailed a letter to the court from the backcountry? This wasn't the only oddity involving Randy's signature and dates. The parks' administrative officer had been astute enough to note that Randy signed a request for travel expense reimbursement for the June preseason training period on July 21, the day he presumably had left his duty station. How could Randy have signed a document in Cedar Grove, which was at least a two- or three-day hike from Bench Lake?

Both signed documents served to keep alive the possibility that Randy had left the mountains—and the theories abounded as to where he might have gone: East India; Japan; Mexico; Argentina; Moab, Utah; Escalante; the Himalayas; the Colorado River; the blue highways; outer space . . .

On August 22, DeLaCruz received word from the profiler with California's Department of Justice. The DOJ's analysis of Randy's personal diaries found him to be an "elevated suicide risk." The profiler made it clear that the assessment was based strictly on the diaries and did not consider any other information.

Other bits of information trickled in through August and September. The letters that Ned Kelleher had sent out to backpackers issued wilderness permits at the time of Randy's disappearance had worked.

More than fifteen individuals contacted the park recounting "strange occurrences," including a large rock slide near Bench Lake around July 21. Kay Edens, the backcountry ranger assigned to finish off the season at the Bench Lake ranger station, was sent to the reported area of the slide, but was unable to locate anything that indicated Randy was caught in the rubble. Cadaver-trained canines had been dispatched to the area during the search, so it was scratched off the list. In another occurrence, a hiker reported speaking with a "strange individual" near the John Muir Trail junction to Paiute Pass. He said he had heard screaming and yelling near his camp. "A short time later, a single male approached his camp and apologized for making all the noise." The hiker "established that this man was camping alone and the yelling he had heard was this individual yelling at himself out of frustration." He felt the man was unstable, and it made him uneasy, plus he found it "peculiar that this individual brought up the topic of the 'lost ranger' and elaborated further, saying that he thought 'the ranger had an accident somewhere and was probably laying out there.' " The man was described as short: approximately 5-foot-4, with dark, straight hair cut around ear level and a dark tan, and thought to be of Native American descent. There wasn't really any way to follow up on this account, other than to tell the hiker to let the park know if he remembered anything else of interest.

DeLaCruz was, however, able to piece together the mystery of the postmarked letter to the Yavaipai County Court while interviewing one of the last people to see Randy.

Trail-crew supervisor Laurie Church had started with the California Conservation Corps in the mid-1980s and, though small in stature, could pulverize granite with a sledgehammer as well as any of the young men who typically make up the crews. The physical labor suited Church, who had been working in the Sequoia and Kings Canyon backcountry for more than a decade.

Church had come to know and respect many of the backcountry rangers, treating them like an extended family and welcoming them into her camp. She recounted to investigators three separate times she had interacted with Randy during the weeks before he disappeared.

When Randy showed up at her crew's White Fork camp on July 11, 1996, she thought he was "in reasonably good spirits." But in hindsight, she reported that certain comments from the evening's discussion were possibly significant. "One was with regards to how long he has been rangering, along the lines of 'I'm finally starting to wonder if it's been worth it.' " This was the first time Randy had ever expressed any dissatisfaction with his job to Church. "Otherwise, he spent a good two hours talking about Peet's Coffee and he seemed all right, maybe a little lonelier than usual."

Before heading back to Bench Lake the next day, Randy invited Church and her crew to come and visit him the following week. Church agreed, because it extended her crew's maintenance range a bit as they repaired the winter storm damage to trails. Randy "hadn't usually pressed us to come visit," said Church, "but I really got the impression that he wanted us to come."

On July 17, Church showed up at the Bench Lake station with a five-person crew. It struck Church as odd that Randy wanted to both host them and share his food, which was rationed carefully for the season. "In most situations we'd camped near him, he'd never really offered us to come and eat with him," she told investigators. During the course of the night, Church mentioned to Randy that she had gotten married in June and that it was difficult being away from her husband for long periods of time. She observed that living and working in the mountains was hard on a relationship. Randy responded, "I know what you mean."

Later, Randy handed Church a book he'd just finished reading that he thought she might enjoy: *Blue Highways* by William Least Heat-Moon. She described it briefly in the interview as being "about a guy that was having marital problems. It's a true-story account of how he dealt with his problems, which was basically that he hit the highway and took an 11,000-mile road trip." This book clued DeLaCruz in to the possibility that Randy could indeed have left the park.

He also found it interesting that Randy had mentioned another book to Church and her crew, apparently what he'd been reading at the time, "about a guy named Everett Ruess," explained Church, "who loved the wilderness and disappeared and was never seen again." DeLaCruz wanted to know more about this story of Everett Ruess.

Toward the end of the interview, Church recounted the last time she saw Randy: July 18, just three days before his disappearance. The trail crew was gearing up to work a section of trail near Taboose Pass, and Randy "seemed a little lonelier than usual," said Church. "I felt drawn to him quite a bit. He's always been a loner, but for some reason I really wanted to spend time with him that morning. We went off and did our maintenance run, and when we returned, the crew packed up their stuff and left and I kind of hung out a little while. I told him I wished I could stay longer, but we had to go back because . . . we needed to break camp and move to another location in the park."

At the moment of departure, Randy handed Church a letter to mail for him when she left the backcountry the following Monday, telling her, "You can bend, fold, spindle, or mutilate it." Then he laughed and said, "Well, maybe not mutilate it." Church, who assumed the letter, addressed to a court, had something to do with the Doug Mantle altercation Randy had told her about, slipped it into her pack.

"That was how Judi got the letter from the court," says DeLaCruz. "Laurie Church had it in her pack for a few days before she left the backcountry and dropped it in a mailbox on July 22—the date of the postmark."

That was one mystery solved, but what about the signature on Randy's reimbursement for travel expenses? It was discovered to be a forgery made by one of the administrative employees in the parks' payroll department. In order for rangers to get these reimbursement checks deposited into their bank accounts as soon as possible, the payroll employee had signed expense forms by holding them up against a sunny window and tracing the signatures from existing documents, a practice Randy had consented to.

Nothing ever came of the strange individual who had been yelling at himself, or any of the other leads generated by hikers who had been in the region. One particularly intriguing rumor, however, had weaved through the ranger ranks and gotten back to DeLaCruz.

At the outset of the search, one team of rangers encountered a man and a woman who'd been in the mountains for nearly a week and were currently hiking over Pinchot Pass on the John Muir Trail. As is common on mountain passes, both groups paused to catch their breath and struck up a conversation. Noting the unusual amount of helicopter activity, the woman asked what was going on, and a ranger showed her the Overdue Hiker flyer that had been posted at all the trailheads. The woman suddenly appeared distressed. Her partner, apparently her husband, said, "Tell them." The woman then explained how, throughout her life, whenever she was in the mountains or wilderness area for any amount of time, she'd have dreams. The man corrected her. "Visions," he said. "She has some psychic capabilities that come out when she's alone and it's quiet. It doesn't happen at home."

The dream, vision, whatever, had been of a man in great distress, trapped and trying desperately to free himself from underneath something—she couldn't tell what. A boulder? A tree? Water? It had startled her from sleep. Now that she had heard there was someone missing in the area, she became genuinely worried. She promised she'd contact a ranger if she had any other episodes.

What made this account different from those of the other psychics who had come out of the woodwork once the story hit the papers was that this woman, if she was being honest, had experienced her vision before she was made aware of Randy's disappearance.

DeLaCruz also learned more about Rusho's book on Everett Ruess,

the true story of Ruess, a 20-year-old nature seeker who vanished without a trace in Utah's Escalante wilderness in November 1934. Randy
and Ruess, though vastly different in age and living in different eras,
shared an intimate bond: they were both artists whose work was inspired by the wilderness. Ruess was a painter who carved woodblocks
and wrote prolifically of his experiences; Randy was a photographer
and writer who used the Sierra as the basis for numerous short stories and editorials and thousands upon thousands of photographs and
pages in his journals.

Ruess didn't realize fame during his lifetime because, it seemed, he
had been unwilling to relinquish his time in the wilderness in order to
market his art. He was more interested in making just enough money
to allow him to return to the wilds. Because of his disappearance, Ruess
became a legend in wilderness lore, and his artwork and writings met
with widespread acclaim.

Rusho's book consists predominantly of Ruess's letters, which were
riddled with dark foreshadowing. In 1931 he wrote: "I intend to do everything possible to broaden my experiences and allow myself to reach
the fullest development. Then, and before physical deterioration ob-
trudes, I shall go on some last wilderness trip, to a place I have known
and loved. I shall not return." The following year, he seemed to put an
exclamation point on this prediction: "And when the time comes to
die, I'll find the wildest, loneliest, most desolate spot there is." In addition to his forlorn musings, Ruess told a friend two years later, "I don't
think you will ever see me again, for I intend to disappear."

Coincidentally or not, *Everett Ruess: A Vagabond for Beauty* was the
last book Randy was reading. Furthermore, he had spoken about the
book or Ruess to nearly every ranger at training that season.

Was Randy thinking of running away from his problems? Perhaps
losing himself? Did he have a premonition of death? Was suicide on
his mind? If his choice of reading material that summer was any indication, one would have to assume that this might have been the case.

DeLaCruz was baffled. But he'd had his own decidedly grounded,
albeit clichéd, premonition. He felt confident that "Time will tell."

◗

THREE WEEKS AFTER the Morgenson SAR was called off, a blue, five-window 1953 GMC longbed—lovingly restored—rolled north on Highway 395 out of Lone Pine, California. Inside, retired National Park Service ranger Alden Nash leaned forward on the steering wheel and looked west at the jagged eastern crest of the High Sierra.

"If he's up there, I don't know if they'll ever find him. Those mountains have a way of keeping secrets," he said matter-of-factly. "Huge metal airplanes crash up there and aren't found for decades. Sometimes never."

Nash had made a hobby of looking for these elusive metal phantoms while patrolling the high country and checking in on the twelve to eighteen (depending on the year's budget) backcountry rangers he supervised in Sequoia and Kings Canyon from 1975 to 1993. His beat had been the backcountry from the Kern River to the San Joaquin River, about 120 air miles north to south and 50 air miles east to west. Nash told most people his job as the Sierra District subdistrict ranger was "the best job in the world"—in part because he was supervising people he described as "the best rangers in the world."

Nash took another long drink from the warm bottle of Gatorade he customarily kept in his truck to rehydrate at the end of a long hike with a heavy pack. The Sierra was no longer Nash's office, but it was still a part of him. ("You can take the ranger out of the Sierra, but not for long," Nash has been known to say.) Cruising along at a steady 55 miles per hour, he shook his head. "Why?"

It was one of Nash's favorite questions, usually timed with a dramatic pause before continuing. "Why would anybody in their right mind put a 50-pound pack on his back and drag himself up and over that?" He directed his gaze toward the looming wall of mountains.

He'd just returned from a week in the high country spent looking for Randy. His unilateral effort had no affiliation with the NPS's official SAR operation. And to his two hiking buddies, it had seemed like any normal Nash death march, albeit with a little more bushwhacking than usual. Only two of seven days in the backcountry were spent on trails—going in and coming out. Otherwise, it was all cross-country—Randy country.

Every once in a while on the hike, Nash had taken off his pack, leaned up against a tree, and rocked back and forth, scratching his back like a bear and providing some temporary relief to a nerve that had been bothering him for years. It was during such a moment that he said to his friends, "Let me know if you smell anything dead."

The statement sounded harsh, especially coming from a man referring to the potential corpse of an old friend, but Nash had spent years desensitizing himself by pulling dead bodies out of these mountains, sometimes in pieces. In addition, Nash, despite his efforts, wasn't entirely convinced that Randy was still *in* the mountains.

Back in the truck, he continued talking to his hiking partners. "The problem they were up against, besides 80 square miles of wilderness, was Randy himself. If he didn't want to be found, he won't be. He knows how these investigations work. He knows how to cover his tracks. Or maybe he . . ."

Nash stopped short of verbalizing suicide but got the message across nonetheless: "Randy wasn't really himself last time I saw him. I wouldn't doubt it if he either made a mistake, or who knows. If he did have an accident, it was probably someplace gnarly.

"But knowing Randy—the Randy I knew, or thought I knew—none of it makes sense. It's a mystery, that's the only thing that's certain right here and now."

The truck headed north, past Manzanar, the Japanese internment camp from World War II. Tumbleweeds blew freely across a grid of dirt roads and cement foundations from long-since-destroyed barracks. It was a sad reminder of what had been located in the Owens Valley in the early 1940s.

"Not one of our country's proudest moments," said Nash, slowing down to a respectful 40 miles per hour as he drove by the camp's ghostly remains. "Officially, nobody ever tried to escape from Manzanar, but as the story goes, teenagers occasionally escaped for a night into the mountains and then escaped right back into the compound

the next day after spending a cold and lonely night. With those godfor-
saken mountains to the west," he said with a wink, "and Death Valley
to the East, they didn't even need fences.

"That's what some people see when they look at that wall of peaks.
It's a fence, a big, scary fence. Not for Randy. For Randy those mountains right there meant freedom."

Nash angled the truck casually onto the shoulder so an oncoming
SUV, passing on a double-yellow line, didn't hit him head-on.

"Freedom from idiots like that," he said.

AS FALL CAME to the high country and the 1996 season drew to a
close, no further clues surfaced, either in the backcountry or via Special Agent DeLaCruz's far-reaching detective work. Randy's Toyota
truck was eventually released as evidence, and George Durkee and his
wife, Paige Meier, volunteered to drive it back to Sedona with Randy's
personal items loaded in back.

But first, Durkee had an end-of-season report to write. He had
always begun his EOS reports with a quote. In this one he immortal-
ized a friend:

> 'We're fine, we're just fucking fine.'—Sandy Graban, Morgenson
> SAR.
>
>   The recommendations included in all of our year-end reports
> could get a bit overwhelming—as well as repetitive. I'd suggest
> that you choose, say, three important issues from all our reports
> and put energy into accomplishing them. I can't think of a single
> specific recommendation we've made in the last 15 years that's
> been acted on. This leads to a certain cynicism . . . (if you're
> reading this, send me an e-mail with code word ZULU.)

Number one on the list of Durkee's recommendations:

>   Close McClure to grazing as a memorial to Randy. I'm quite
> serious about this. Randy Coffman and Rob already talked

this over and RC wasn't too keen on it. Debbie Bird won't even discuss it. I firmly believe it's the only meaningful act we can do to honor someone who worked here 30 years without the least bit of official recognition from the park. Randy worked at McClure for 8 seasons. In EVERY ONE of his year-end reports, he recommended closing it to stock use.

IN A CONTINUED EFFORT to locate Randy's body, Subdistrict Ranger Cindy Purcell took to the sky for two hours and fifteen minutes in the park helicopter after one of the winter's first snowfalls in December. It was a recommendation from Durkee, who had discovered that animal tracks in the snow often lead to dead bodies—animal or human. He'd watched coyotes converge on deer killed in avalanches on numerous occasions while performing snow surveys, coincidentally, with Randy.

Purcell spotted one distinct set of animal tracks in the vicinity of Bench Lake, but "no disturbances in the snow."

If the mountains knew where Randy was, they weren't telling.

CHAPTER THIRTEEN

SERMON ON THE MOUNT

Who will speak for the trees?
—*Chief Justice William O. Douglas, 1972*

If I am among the mountains, yet in a sour mood or
with my thoughts elsewhere I hear not their voices—feel
not their Presence and Forces.

—*Randy Morgenson, Nepal, 1969*

IN JUNE 1997, about a year after she last saw Randy in the backcountry
at Bench Lake, Kings Canyon Sierra Crest Subdistrict Ranger Cindy
Purcell stood before a group of his peers at the Cedar Grove visitor
center. It was her job—"honor," as she put it—to begin the memorial
service for Randy Morgenson.

Most of Randy's closest ranger friends were there, but some, including Judi, couldn't bring themselves to attend a memorial without a
body or, for that matter, an official declaration of death. Nevertheless,
Purcell recounted those four "special" days in which she had gotten
acquainted with the ranger who would become a mystery.

"Randy was as complex an individual as any one of us," said Purcell. "His dreams had conflicts. His ghosts were big and scary. But his
spirit was so full of joy and love that he could overcome his doubts

and move through them." The summer season of 1996 "was a time for Randy to go through a great deal of soul-searching, some of which he shared with me.

"There was a great deal of personal strength in this man. He told me that this was the strength he got from the Sierra. He loved this place with his total being. During the winter of 1995, he toyed with the idea of not returning for another summer, but the mountains called him back.

"This is the time to remember the events of last summer that left each one of us empty and questioning.

"What drew him away from his ranger station on that day in July?

"He was in 'ranger mode.' It was time to patrol. Perhaps he was called by some 12,000-foot peak, or a favorite place visited too long ago needed revisiting. . . . We shall never stop asking, 'Randy where are you?' "

In addition to leading the memorial service, Purcell had taken it upon herself to research any available awards in the National Park Service that might honor Randy. She discovered the Truman James Memorial Award (Truman James was a seasonal SEKI employee killed in a tree-trimming accident), created to recognize "deserving seasonal employees . . . who exhibit special environmental awareness through conservation efforts, preservation efforts, educational enlightenment of the Park's visitors, or special concern to visitor safety." Randy qualified for all of the above.

Purcell's proposal for the award sung Randy's praises. "He could identify any plant or wildflower by genus and species with an enthusiasm that encompassed him," she wrote. He possessed "a very special environmental awareness . . . specifically for the fragile meadow areas he loved. He maintained undying energy to lobby for the closure of meadows where he monitored irreparable damage to fragile species . . . energetic in the education of the wilderness visitor to support minimum impact ethics and encourage the visitor to assume responsibility for protecting the parks . . . a base of feelings strong enough to be felt by those of us fortunate enough to have worked with this outstanding individual."

The award—Randy's name on a plaque displayed at the Ash Mountain headquarters—was granted shortly before his memorial.

Purcell had also spent months combing through Randy's logbooks in preparation for the service, pulling out hundreds of worthy passages that she had edited down to thirty-five, which were handed out among memorial attendees to read aloud.

There were chuckles from the crowd during some of the readings, such as the passage recounting the time Randy had "a long and sunny, very congenial and mutually rewarding lunch with a chipmunk. He is a little timid, understandably enough considering my relative size, and reserved his comradeship for a day when there was only one of me."

Or one of the hundreds of times his radio had died, and the park helicopter flew in to bring him "a new radio, right? Wrong. But a large and weighty burlap sack was unloaded, so I sent my 2-watt radio back out. Alas, the burlap sack only contained a bottle of wine and more chuck roast and rib steak than I could eat in a week. So here I sit with half a cow wrapped in my sleeping bag and no way to call for any of my compadres to come help me with it."

One of the last passages read had been written by Randy during his first season at Rae Lakes, more than thirty years before. Even then, as a young man, he'd captured the subtle nuances of autumn in the high country—the trigger for backcountry ranger melancholy. "An extremely clear, perfectly cloudless day," he wrote. "Fall has definitely come to the High Country. The air is clearer and cooler, night-time temperatures are close to freezing, the sky is much bluer in the afternoon, very deep and dark in the east, blending into a lighter color in the west—and the caressing summer breezes have become gusty afternoon winds. The whole atmosphere seems quieter, so the animal sounds seem louder and more clear, and the wind more hollow. Tis a beautiful time of year, but a somewhat sad one, for it brings the end of the Sierra season."

"After that quote," remembers Nina Weisman, "you could have heard a pin drop."

Following the memorial, Weisman hiked to her backcountry ranger station at Bearpaw Meadow, where the first thing she did was post the quote she had read in a place of honor—above the toilet paper roll in her ranger station's "facilities."

"Sitting on a rock for the noon radio check, halfway down the South Fork, I feel no questions, no troubles, just a great oneness with all welling up inside me. This moment is all that is, all that ever will be. Memories can never equal the experience, and at best we can only attempt to visualize the future. The best we can do is absorb the most possible from Great Moments Like These."

"Randy," she says, "would have appreciated the humor."

Weisman then took out the laminated Missing Ranger bulletin she'd had on the window of her cabin after the search had ended the season before. It urged hikers and climbers to "Please be on the look-out for an abandoned camp or any scattered pieces of clothing, back-pack or its contents or human remains. Boulder fields, bases of cliffs, couloirs, and lake shores are areas to be alert in." Now, at the start of her second season as the Bearpaw Meadow ranger, she reposted it and intended to repost it each season, no matter where she was stationed, until the mystery was solved. Though she hadn't been part of the SAR itself, she would see to it that backpackers passing through her section of Sequoia and Kings Canyon knew of Randy's disappearance. It was her way of making sure Randy's ghost marched into the ranks of wil-derness legends, "not to be confused with those who earned the dis-tinction simply because they walked into the wilderness one day and never came back," she explains. "Randy Morgenson is a legend because of the life he led here in the Sierra." His disappearance was merely the catalyst that brought into focus the legendary and selfless acts he'd per-formed—which Weisman made sure were known to anybody asking about "that missing ranger."

Virtually all the backcountry rangers, and some from the front-country, did their part to keep Randy's spirit alive. In the back of the class during law enforcement training, they quietly discussed the places they intended to search once they were flown to their stations. At night, they gathered around bottles of wine and told stories of "Morgensonia."

At their stations, they pored over the dusty files and logbooks, seeking tangible pieces of his ghost in the fading, mouse-chewed archives. The backcountry rangers called these words written by Randy Morgenson "The Gospel."

ALMOST IMMEDIATELY after the search was called off, Chief Ranger Debbie Bird had recommended that Judi file for the Public Safety Officers' Benefit Program.

Judi, whose life had been in turmoil for three years, was encouraged to pursue a death benefit for Randy, whom she had been in the process of divorcing. But it didn't sound right to her; it sounded as if she was unwilling to stay with the man, but she certainly was willing to take money to honor his death. She was on the fence until Durkee spoke to her. "The point I tried to make to Judi," he says, "was that Randy didn't want the divorce—his letter to the court proved that." He told her that, whether she liked it or not, she was still married to Randy. Durkee also reminded her that Randy hadn't felt particularly appreciated for his twenty-eight years of service and this benefit, however late, was a symbol of appreciation. "Judi, Randy would want you to have this," Durkee told her. "He knew the pain he caused you."

Judi consented, and in doing so opened up a can of worms. While friends and colleagues kept Randy's spirit alive, Judi found it exceedingly difficult to prove his passing. She knew Randy was dead, just as she had known when he was having an affair. "A woman just knows these things," she says.

Admittedly, her heart skipped a beat every time she saw a man with a dark beard approaching her on the street. Numerous times, she'd envisioned Randy knocking on the door, and she would hug him and kiss him . . . and then, with a bit of Hollywood drama, she'd slap him and beat on his chest and hug him again. Every single time the phone rang, she wondered if maybe, just maybe, it was Randy, ready to come home. She was trapped in the cruel and inescapable whirlpool of not knowing. Her life was at a spinning standstill. Even if she had the desire, she couldn't go forward with her own life. How could she, in good conscience, move forward in a new relationship without closure? On the flip side, she was gearing up to fight for a benefit based on the belief that there *was* closure—that Randy was, with certainty, dead. It was maddening. More than once, when she didn't know if she could go on with the claim, Durkee stepped in.

"Thank god for George and Paige," says Judi.

Durkee had made it a priority to see that Judi Morgenson received the Public Safety Officers' Benefit of $100,000 from the U.S. Department of Justice, not only for Judi but also for Randy. With this purpose in mind, Chief Ranger Debbie Bird essentially gave Durkee and Judi the key to the parks' administrative offices. If there was anything the parks could do, she would try to facilitate it.

Two basic elements are needed for benefit eligibility. One: a body. Two: proof that the officer has been killed while on duty. Without number one, number two cancels out by default. But Bird did some research and found that "proof of death" could suffice in lieu of a body. This was possible if the director of the Park Service or the secretary of the Department of the Interior declared Randy dead.

With the help of Bird and Durkee, the request for an official proof of death was made. Now it was time to wait.

MORE THAN A YEAR after the original search for Randy was called off, there still was no ruling on his death. Judi did, however, return to the Sierra. She'd been invited by Debbie Bird to Bench Lake, an opportunity to meet some of the rangers who had searched for her husband and to pay her last respects as a follow-up search did a final sweep of the area for Randy's remains.

When Judi boarded the park helicopter on September 22, 1997, she was both nervous and oddly at ease. She felt, in a way, as if she was coming home. As they flew above LeConte Canyon and toward Lake Basin, the helicopter decreased altitude. Looking down at the deep blue waters of the Dumbbell Lakes made Judi gasp, an abrupt physical reaction that gave her chills.

She'd had that "vivid dream" shortly after Randy's disappearance of him floating with a backpack on at the bottom of a lake. "After that dream," says Judi, "I always thought he was under water. That vision never let up." Once they landed at Bench Lake, Judi relayed the information to Eric Morey, the incident commander, who assigned a team to thoroughly check the lake once again. "They didn't find anything," says Judi. "They sent out technical climbers from Yosemite to search the cliffs they couldn't access the year before. Nothing. Not a clue. But he was out there. I could feel it."

Before the current search began, Cindy Purcell had asked all the rangers to contribute their theories. One ranger wrote, "Have been scratching my head on this search . . . it does seem to be in the 'rolling over rocks' mode." The ranger then suggested eliminating areas that had been well-searched: "Example—Explorer Pass is a narrow chute. If he had fallen there, chances are we would have found him." The suicide theory was still alive: "Look closely in areas where he may have jumped, particularly on the north side of Arrow and Marion Peaks . . . most likely scenarios: a total immobilizing injury . . . or a swan dive. Other than this, there would only be bone fragments left, with his pack and camping gear being the most likely items to be seen."

Five days of intense searching by twenty ground searchers uncovered not a single clue.

JUDI FLEW HOME to Sedona and dived into her art. She decided to turn the garage into a studio. This would mean selling Randy's 1932 Ford Model B five-window coupe, which he'd held on to since he'd bought it in high school. Before he left for the summer, he had told Judi to "sell it if you can get a good price." Now, she decided, maybe it was time . . .

After Randy's disappearance, Judi realized that she was the keeper of many of the Morgenson family treasures. Randy's brother, Larry, would call Judi to "check in" on her, but the conversations always ended with him requesting some piece of his parents' property that had been left to his brother. Some items Judi didn't mind letting go of, but she felt a duty to protect the memories and to honor Dana and Esther as she knew her husband would have. Eventually, she decided that Larry was more interested in the property than in any updates about her or his brother. After a few years, she asked him to stop calling.

Beyond his parents' books and keepsakes, there were other things belonging to Randy that she knew he wouldn't have wanted to sit unused in boxes. In particular, his camera equipment. Stuart Scofield was the one person Judi knew Randy would have wanted to go through his photography treasures.

Scofield arrived in Sedona uncertain about how he would react when he went through Randy's personal belongings. He hugged Judi, and on cue, tears welled up in her eyes. She explained how tears came easily anytime she encountered someone associated with Randy. After they caught up, Scofield was left alone in the first bedroom off the house's entrance, which had served as Randy's main library. An entire wall was filled with books, and boxes of slides, prints, and equipment were stacked around the room. Judi had also directed Scofield to a dresser filled with photography odds and ends.

Scofield admits that before he opened one of those dresser drawers, he had not been prone to metaphysical thinking. That changed when he pulled out a cardboard box about the size of a hiking boot shoe box that was filled with random items such as a flash, battery pack, camera batteries, and lens cleaner. It looked like a neatly arranged care package intended for a photographer. Sitting atop the assortment, centered in the box, was a small pamphlet "like you'd get in Sunday school," says Scofield, "and on the cover of this little pamphlet was Sermon on the Mount, from the Bible. It totally threw me backwards—back in time to Randy's Sermon at Mineral King."

It was the late 1980s, when Scofield and Randy had taught a photography workshop called Wilderness Landscapes. They'd camped at the Potwisha campground inside the parks' Ash Mountain entrance, and woke with the students early one morning to make it to the nearby Mineral King Valley while the light was still right for photography.

The night before had been a sky show of "fantastic thunderstorms," and there was still a lot of moisture in the air. Randy had planned to inspire the students by telling them how Walt Disney almost turned the Mineral King Valley into a huge resort. This was the type of thing Randy and Scofield did in their workshops, which weren't just about how to take photographs. They were, according to Scofield, "about how you decide who you are out in the world as a photographer, and if you're going to photograph the landscape, how you develop a rapport with the wilds."

That misty morning, Scofield, Randy, and a dozen students gathered at the trailhead and focused their attention on the stunning Mineral King Valley. Randy discreetly climbed up onto a rocky knoll and began reading what Scofield described as "amazingly powerful passages" that Randy had prepared about this, the site of one of the environmental movement's greatest battles: *Sierra Club v. Morton*—the 1972 Supreme Court case widely referred to as *Mineral King versus Disney*.

Fog rolled in and around Randy as he spoke from atop the granite podium, sometimes shrouding him almost entirely from the students. But his voice, reading the powerful words of Chief Justice William O. Douglas, was strong and steady:

> So it should be as respects valleys, alpine meadows, rivers, lakes, estuaries, beaches, ridges, groves of trees, swampland, or even air that feels the destructive pressures of modern technology and modern life. The river, for example, is the living symbol of all the life it sustains or nourishes—fish, aquatic insects, water ouzels, otter, fisher, deer, elk, bear, and all other animals, including man, who are dependent on it or who enjoy it for its sight, its sound, or its life. The river as plaintiff speaks for the ecological unit of life that is part of it. Those people who have a meaningful relation to that body of water—whether it be a fisherman, a canoeist, a zoologist, or a logger—must be able to speak for the values which the river represents and which are threatened with destruction.

"Who," asked Randy with a dramatic pause before quoting the rest of Douglas's most famous line, "will speak for the trees?"

"When he finished, the students cheered," says Scofield. "The whole scene had taken on these epic proportions—the jagged Sierra crest farther up the valley coming in and out of focus through the mist. It was incredibly moving." And a mini history lesson: most of the students didn't know that Mineral King had been the case that raised the question of rights for inanimate objects. Scofield, who has taught workshops for more than twenty years, says he has never seen an instructor move an audience the way Randy did that day. "It was like Jesus on the mountain," explains Scofield. "Powerful weather, amazing beauty, the fog, the crest. It just all came together."

And it all came back to him now, vividly, as he sat among Randy's belongings. Ever since that day at Mineral King, Scofield had referred to Randy's speech as "Randy's Sermon on the Mount." He was suddenly overcome by the unshakable conviction that this care package was not just for any photographer—it was tailor-made for him. "There was no question in my mind," says Scofield. "It was a message from Randy."

Later, Judi showed Scofield some of the endless files that Randy had kept on virtually every aspect of his life—files that held documents from grammar school, high school, college, the Peace Corps, his correspondence with Wallace Stegner and Ansel Adams, his ranger training, his mother and father, and an extensive documentation of his photography, including the workshops he'd taught.

When Scofield saw those files, he was further convinced that the box was indeed a message from Randy. "If he had wanted to file that Sermon on the Mount pamphlet away as part of his history, Randy kept files of things like that," he says. "That is where he would have put it. If you assume for a minute that he did want me to find it, he would have known that Judi would have had me go through his camera equipment, but may not have gone through his files. And sure enough, Judi invited me out to do just that. Randy knew. He was thinking about either disappearing or taking his life when he packed that box. He wanted me to have something, a piece of him.

"It worked, all right—that little pamphlet hit me like a ton of bricks."

Scofield shared with Judi the story of Randy's Sermon on the Mount, but he wasn't entirely forthcoming about his belief that the pamphlet had underlying meanings. Judi was grieving, and he didn't want to add to her burden—certainly not that Randy may have been planning his disappearance or his death.

Scofield did, however, have one other theory, though it was "out there." He considered it possible that Randy might have put this box together simply because "he sensed something might happen to him." Perhaps if Scofield had brought up that theory, Judi might have told him the "message" she herself had gotten from Randy, a message delivered, appropriately, within the pages of a book—the novel he'd given her before he left for the mountains.

I Heard the Owl Call My Name had sat on her nightstand until the third or fourth day of the search. She had by that time spent her anger and accepted, or more accurately "knew," that something had happened. She describes the novel, but only after explaining that Randy always read slowly. "Because he loved the language, he'd savor the language," says Judi. "He didn't do anything quickly. He felt you'd miss things if you went too fast. We were open to things, not the mystical New Age stuff here in Sedona, but being in the mountains and spending that much time by yourself . . . you can tell by his writings he was really in tune with the cycles of nature. You get into those cycles and you become very aware of them, especially when your mind is quiet. It's very Zen. Everything slows down and you can hear things. You can hear yourself."

The novel was, in Judi's words, about "a young minister who was sent by a bishop to British Columbia to live with some native Indians whose tribe was slowly disappearing, losing its old ways and its people. He went there to help them, and in the process learned a lot about life and death and how to accept death. Toward the end, he heard the owl call his name, which was a myth that they had in their tribe. What it meant was that he was going to die. The minister didn't know it but he had terminal cancer."

"Subconsciously," says Judi, "maybe Randy knew. When you're

tapped into the kind of emotional pain he was going through, you are really in tune with other things in nature. The things around you, you just know things—maybe even premonitions. Shortly after reading that book, I had the dream about him being in that lake, under water. So, you get tuned in.

"I think Randy might have had that feeling. Maybe he felt that. Maybe the mountains called his name?"

Some would consider this metaphysical hogwash. In fact, after some time had passed, Judi herself began to discount the possibility that Randy had tuned in to something, remembering that Randy knew she liked owls and had perhaps seen the book and thought of her. Nothing more.

Judi might have thought differently if she had known about Randy's patrol the season before he disappeared. On September 17, 1995, Randy had hiked from LeConte Canyon down the White Fork to Woods Creek Crossing and then camped in Paradise Valley. Forest fires were raging in the parks. It was raining ash and smelled of acrid smoke. The sun barely cut through the darkness, and then, come nightfall, Randy's headlamp beam created the same effect as car headlights during a snowstorm. Because of the ash, Randy slept inside his tent, which was unusual.

In the morning, Randy was awakened by an owl calling. It was a surreal day, with silver-and-white ash so thick on the ground it looked like snowfall. He wrote in his logbook, "A great horned owl calling in the wee hours before dawn; eerie smoky dimness, and the owl calling—Paradise."

Coincidence? Randy might have said, "Only the owl knows for certain."

But what is known is that great horned owls are common in the parks. It stands to reason, then, that Randy would have mentioned great horned owls calling dozens of times over the years. According to his station logbooks, however, during the entire course of his ranger career, he'd never mentioned hearing the great horned owl—or any owl, for that matter—"calling," and he'd documented literally thousands of other species, describing in detail their songs, voices, music. Never an owl. Never its call.

Was it a coincidence, then, that nine months later, he gave his wife *I Heard the Owl Call My Name* as a parting gift on the last day she ever saw him? Did it strike as anything less than odd that of the hundreds of books lined up on his bookshelves, of the thousands in the bookstores he frequented, he would pick a book whose main message is that of a man's ability to sense his own coming death—if he listens.

Randy had heard an owl calling and described it as "Paradise." Did he hear anything else? Was there a message? In the wilderness, nobody listened more attentively than Randy.

ON JANUARY 14, 1998, the director of personnel policy and the so-licitor of the National Park Service informed Judi that they had com-pleted their review and "found that circumstances surrounding the disappearance of Ranger Morgenson support the presumption that he is no longer alive."

A year and a half after she'd submitted the request, Judi had legal proof of Randy's passing. George Durkee put together what he de-scribed as a "bombproof" report "proving" that Randy had been killed while on patrol in the backcountry. It included an inch-thick investi-gation report supplied by Al DeLaCruz, the declaration of death, and the delayed death certificate. In the beginning of March, Park Super-intendent Michael Tollefson sent the claim on Judi's behalf to the U.S. Department of Justice, Office of Justice Programs, Bureau of Justice Assistance.

Seven months later, her request was denied in a letter that read, in part: "The evidence submitted on your behalf is insufficient to con-clude that your husband's death was the result of a line of duty . . . this office recommends that you provide additional evidence that demon-strates . . . a line of duty death. We regret the loss of your husband, Ranger Morgenson. If you have any questions, please call at . . ."

Judi resubmitted a beefed-up version. Five months later, she re-ceived an official denial that stated "while the PSOB Act requires this determination, such action does not diminish Park Ranger James Randall Morgenson's distinguished record of public service." Attached was a two-page, play-by-play reasoning for the decision, including "Ranger Morgenson's wallet and duty weapon were recovered from his duty station."

Further, the denial letter stated that the evidence submitted by Judi "is insufficient to conclude that Ranger Morgenson was engaged in duties which he was authorized or required to perform as a park ranger at the time of his death. Ranger Morgenson's body was not found resulting in the issuance of a Delayed Registration of Death. . . . The record, as it currently stands, provides very little for this office to make a favorable determination."

The letter then referenced a case from 1985, *Tafoya v. United States,* whose ruling "prohibits the use of conjecture and speculation as evidence or fact when it concerns the cause of an officer's death."

In conclusion, "Accordingly, Ranger Morgenson's survivor is ineligible to receive the benefit authorized by the Act."

When Durkee read the letter, he began fuming. He could picture a bunch of guys sitting in an office in D.C., utterly clueless about the Sierra terrain. "The fact that Randy hadn't been found would make no sense to them and was probably reason enough to deny the claim," Durkee told Judi. "The only way they'll pay is if we find him dead, in his uniform, radio in hand."

In essence, the Bureau of Justice Assistance left Judi no other choice but to sue, which would be a battlefield strewn with coils of red tape, endless paperwork, and mountainous legal expenses that would likely eat up a good portion of any benefit she'd eventually get. But she couldn't look behind the injustice. How could a backcountry ranger, whose job it was to patrol—alone—the most rugged terrain in the country's most treasured wilderness lands, be denied the benefits he thought existed before committing himself to the wilds?

Of course, there still remained the theory that Randy had staged his own disappearance. What if she fought the battle and Randy happened to show up from Mexico with a sombrero and a "Sorry."

"Impossible," would be Judi's response to such a line of questioning in a court of law.

"Why is it impossible, Mrs. Morgenson?" the opposing counsel would probe.

"Because I just know" would be a difficult response to prove in court, especially with $100,000 on the line.

Then there was the veiled suggestion that because Randy's duty weapon hadn't been with him, he wasn't really on duty.

No backcountry ranger was required to carry a sidearm on cross-country routes—or so Judi had been told. But an astute attorney would point to NPS-9, the 2-inch-thick Law Enforcement Policy and Guideline Handbook that Randy and all law-enforcement commissioned rangers were required to know inside out. As early as 1989, NPS-9 stated, in Section II, Chapter 3, page 2, that "commissioned rangers on backcountry patrol shall keep their defensive equipment readily accessible or, at their discretion, may wear defensive equipment on the uniform belt." The "minimum" defensive equipment to be worn, as stated in NPS-9, includes "handcuffs and case, a 4-inch barrel revolver with holster, and spare ammunition with carriers."

And on the next page: "Commissioned rangers assigned to duties without law enforcement responsibilities shall follow local park policy."

Any future court case would be a matter of interpretation of the law and would no doubt get ugly, according to Durkee, who was working closely with two different legal firms representing Judi—both of which had been recommended by the Fraternal Order of Police. The first firm had to drop the case for economic reasons when the Department of Justice made it clear that DOJ law prohibits representation on a contingency basis. Attorney Kenton Komadina of Yen, Pilch, & Komadina in Phoenix, Arizona, reviewed the case and agreed to take on Judi Morgenson as a pro bono client in late 2000.

Shortly thereafter, Durkee put together a comprehensive list of points for Komadina to review, the first being that PSOB law did not include any guidance for situations in which an officer's body is not found. The law had apparently been written with urban, or at least civilized, settings in mind, settings in which a peace officer killed in the line of duty would certainly not, for example, be covered indefinitely by a rock slide or an avalanche, or be missing in such a vast landscape as to be in effect erased.

Debbie Bird has studied the act scrupulously and feels that "the intent of Congress was that someone like Judi would be eligible for the benefit." But the reality was that "the law governing the eligibility is unforgivably vague."

Beyond the benefit itself, Judi had a cause to stand up for: she wanted Randy to be acknowledged for his years of service. Though his life as a ranger had been a strain on their relationship, she'd always respected his tenacious dedication to the wilderness. That he had died while on patrol and was not being officially recognized angered her. Furthermore, there were other rangers who might benefit in the future from the legal precedent this case could set.

Judi appealed the decision in March 2000. Subpoenas were scheduled to be sent out in early January 2001 for a hearing date slated for January 25, 2001. Komadina had a dream team of witnesses lined up to testify, including Randy Coffman, Cindy Purcell, Debbie Bird, and George Durkee. But there was a problem: the hearing was in Phoenix and the witnesses' travel expenses would not be covered by the NPS, which left Judi Morgenson unable to afford what amounted to thousands of dollars. Durkee volunteered to pay his own way, but his testimony alone wasn't worth the gamble.

A last-minute motion delayed the hearing and thus bought some time to consider different options. Though this was potentially a longer road, Komadina was optimistic that a more political route might be the best way to influence the DOJ to reverse its decision.

CHAPTER FOURTEEN

THE DEVIL'S ADVOCATES

Here, where death waited behind each tree, he had made friends with loneliness, with death and deprivation, and, solidly against his back had stood the wall of his faith.

—*Margaret Craven,* I Heard the Owl Call My Name

Am starting to feel a little sad at the prospect of leaving this fine place and this fine living, for the days become finer the deeper we edge into autumn. But there are other good things ahead and nothing remains forever, not even these eternal hills.

—*Randy Morgenson, Crabtree Meadow, 1974*

IT WAS LATE MORNING on July 14, 2001, when backcountry ranger Nina Weisman was packing a backpack to hike away from her duty station at Bearpaw Meadow. The night before, she'd received news of a death in her family, and she was anxious to get on the trail. Her gear laid out on the picnic table in front of the A-frame cabin, she glanced at the sun-faded, barely legible Missing Ranger bulletin she had pledged to keep posted until Randy's mystery was solved. Five years had passed.

As she stood to leave, she had an overwhelming sense that "We're never going to find Randy." So strong was this conviction, she un-shouldered her pack, unlocked the cabin door, went back inside, and took down the bulletin she had taped to the window. Then she headed down the High Sierra Trail to attend a funeral.

On that same morning, 32-year-old California Conservation Corps (CCC) supervisor Peter Martinez from Los Angeles was on a back-packing trip with three young corps members—Evan Ramsey, 19, from Nevada City, California; Mike Noltner, 20, from Middleton, Wisconsin; and Gretchen Haney, 20, from Olympia, Washington—to a little-traveled spot called Window Peak Lake on their weekend off. After camping out the night before, the group worked its way around the lake to its inlet, where Martinez headed up a loose and rocky knoll on the east side and above the creek. The other three bushwhacked up the same gorge where Ranger Bob Kenan had come face-to-face with two mountain lions more than a decade earlier.

These CCC members were part of a special backcountry trail-building unit based near Woods Creek Crossing in Kings Canyon. They were a typical group of strong, motivated youths between the ages of 18 and 23 who, after completing the Corpsmember Orientation Motivation Education Training (COMET) and other specialized train-ing, had hiked into the Kings Canyon backcountry with everything they'd need for their summer job in a backpack. For most, it was the biggest adventure of their young lives.

Modeled after, but not to be confused with, the Civilian Conserva-tion Corps created by Franklin Roosevelt in 1933, the modern CCC was formed in 1976 by California governor Jerry Brown, who saw the program as "a combination Jesuit seminary, Israeli kibbutz, and Marine Corps boot camp." In 1979 the CCC's director, former Green Beret B. T. Collins, coined the still-standing CCC motto: "Hard work, low pay, and miserable conditions." Yet in spite of the grueling work, most of the corps members spent days off exploring the mountains rather than relaxing.

Corp members were warned that these mountains are no joke. "People get lost here," NPS trail-crew leader Cameron Aveson told them, "and people die. There's a backcountry ranger who got into trouble out here somewhere in 1996 and they still haven't found him." The camp cook, NPS employee Kris Thornsbury, had been working with trail crews since 1997. She perpetuated the mystery of Randy Morgenson and always reminded the young and fearless men and women of the corps, before striking out on their weekend adventures, to have fun but be careful.

About the same time Weisman left her cabin on the other side of the park, Ramsey was zigzagging back and forth across the creek in the "mountain lion" gorge, moving up as the pools and waterfalls allowed. Once he was above the thickets of willows, areas of grass and wildflowers appeared, veins of green in the granite mountainside. A few hundred yards above the lake, he observed a weathered-looking backpack on the left side of the creek, a couple of feet out of the water. A coffee mug and a pair of green shorts were within a few feet of the torn, sunbleached pack. Ramsey had been instructed by Martinez to collect any trash found in the backcountry, so he put the mug and shorts inside the pack, along with a water bottle that Haney had found lower in the drainage. They stashed the tattered pack and continued their hike.

After climbing Pyramid Peak, the group retrieved the pack and camped on a flat bench above the gorge. Following a leisurely start the next morning, Martinez took the same route as before, over the knoll on the east side of the creek, while the others went down the more difficult ravine. Around 11 A.M., Mike Noltner discovered a hiking boot in the water at the edge of the creek. Upon closer examination, he let out a yell.

Protruding from the hiking boot was a leg bone. Inside the boot, the perfectly preserved skeleton of a foot.

Martinez scrambled down into the gorge and placed the boot and bones in a plastic bag. They looked deeper inside the old, torn pack they had been carrying back to camp to discard as trash and found a fuel bottle with the initials "S. D." on it.

They built a rock cairn to mark the spot where they'd found the boot, then combed the ravine in a slow descent, hoping to find something else, yet dreading it at the same time. By the time they reached the lake, they'd picked up a thermometer, a granola bar wrapper, what they believed to be a piece of a skull, a large leg bone, a fragment of a pelvic bone, and a bottle of sunscreen. They put everything in the plastic bag.

Shaken by the morbid discovery and anxious to report it, they rushed back to camp. At 2:15 P.M. Martinez showed the pack and its contents to Kris Thornsbury, who noticed that the green shorts looked like Department of Interior issue and deduced, with Martinez, that the gear might be that of a ranger. Thornsbury tried to contact the closest ranger, Kay Edens at Rae Lakes, without luck. She then contacted the Kings Canyon dispatcher.

Subdistrict Ranger Scott Wanek was in the Cedar Grove Visitor Center when a radio behind the counter crackled to life. He'd stopped in briefly and was listening with only one ear when he heard "I think we found Randy Morgenson."

The park interpreter working behind the counter picked up the radio and handed it to Wanek, saying, "You probably want to take this." Wanek quickly suggested that Thornsbury not repeat what she'd just said and advised her that he was en route to her location via the park helicopter.

Wanek then spoke with Chief Ranger Debbie Bird, and an hour later he and the current Sierra Crest subdistrict ranger, Debbie Brenchley, flew to the trail-crew camp, viewed the items, and interviewed the CCC members. Both Brenchley and Wanek were familiar with the Morgenson case: Wanek had been part of the original search effort and Brenchley had reviewed the 4-inch-thick file documenting the effort. As they looked over the collected items, three "potential clues" jumped out at them. The pack matched the description of Randy's, a blue Dana Design, as did the hiking boot—a Merrell. The initials on the fuel bottle, S. D., stood for "Sierra District," which had been the backcountry district involved in the Morgenson search.

One of the most enigmatic clues was the waist belt of the backpack. It was "attached when we saw it," says Brenchley. "The CCC member who found it confirmed without a doubt that he had found it that way, meaning the individual was almost certainly wearing it at the time of death."

Chief Ranger Bird contacted Special Agent Al DeLaCruz to alert him that human remains, believed to be those of Randy Morgenson, had been discovered in the backcountry.

Both DeLaCruz and Bird had stayed in contact with Judi Morgenson over the past five years and understood the emotional stress she'd been under. "A lack of closure is sometimes worse than knowing, because the wounds never heal, they just stay open for years," says DeLaCruz, who had quietly followed up on two John Does fitting Randy's description since his disappearance—one deceased and one with amnesia. Now, DeLaCruz didn't want to contact Judi unless they were absolutely certain the remains were those of her missing husband.

EARLY THE NEXT MORNING, a cadre of rangers, led by DeLaCruz, were flown to the shore of Window Peak Lake. DeLaCruz, who had investigated body recoveries in the past, organized a carefully orchestrated event. He had enlisted the help of proven frontcountry law enforcement rangers to perform such tasks as photographing the scene, taking measurements to document the remains, and collecting remains. DeLaCruz had thought it would be too close to home for the backcountry rangers, most of whom had taken part in the Morgenson SAR and were Randy's friends, to join the investigation, but at Debbie Bird's suggestion, he requested their assistance. Nobody turned down the request. Bird had felt that their inclusion might be helpful in bringing closure. For the backcountry rangers, closure was only half of it. They wanted to understand exactly what had happened.

Two search dogs aided them as they began the morbid process of searching the gorge that drained into the lake.

By noon, the site was speckled with evidence tape, each swatch of yellow marking either a piece of equipment or human remains. DeLaCruz was most interested in a gravelly, flat area at the top of the gorge, not far above a series of waterfalls. There, on the eastern edge of the creek, embedded slightly in sandy, wet soil next to a patch of grass, was a park-issue Motorola MT1000 radio—the uppermost piece of evidence. The radio was switched on, "which meant Randy had either been monitoring radio traffic or attempting to transmit," says Bob Kenan. "That much was immediately certain."

The minute Kenan was flown in, he recognized the gorge—not only from his mountain lion encounter but also because he had been part of three ground teams and two dog teams that searched this drainage, Segment M, on two separate occasions. He was also familiar with the exact location where the radio was found, because it was where he—and all the rangers—crossed the creek when traveling from Bench Lake over Explorer Col and south toward the John Muir Trail. He had used the same route during the SAR to get from the upper part of the drainage down to the lake. Apparently, it was the route Randy had followed as well.

Here, the gorge's walls mellowed into gentle banks of broken granite and gravel sloping inward to a wide section of the creek where grasses and mountain flowers made a living in patches of silt and rocky soil. The creek was either an easy wade or a no-brainer rock hop, depending on the water's level, which, by reading the water marks on the rocks, Kenan confirmed "never got more than a couple of feet deep." It was a cosmic spot where a tired ranger could lean back against his pack, look at the flowers, and have lunch while listening to the roar of the falls beginning 50 feet downstream.

It appeared that the radio marked the spot where Randy had met his end, which utterly and completely perplexed Kenan. If his memory was serving him, he and his team had done a thorough job searching the location during the SAR. "And," he thought to himself, "there were dogs." His memory had not failed him. After searching the area on July 31, 1996, his team's debrief form stated: "Thoroughly covered area; would not go back into this area." If Randy had been right there where Kenan always crossed the creek, how could he have missed seeing at least *something*?

Another ranger on the scene, Dave Gordon, had also searched this drainage early on in the SAR with Laurie Church. Gordon told

Durkee he had been assigned this segment, and though his memory, like Kenan's, was spotty, he felt certain he would have searched along this creek since water sources are magnets for people lost or injured in the wilderness. Gordon suspected that the reason he had not searched this gorge was because of snow or ice.

Durkee was photographing and recording the GPS coordinates of locations where evidence was collected—the radio at "370942E x 4084427N." As he worked, he formulated in his mind what could have occurred. The steep slopes that rose for more than 1,000 vertical feet above both sides of the creek offered a potential clue. He theorized that an avalanche might have deposited a large amount of snow in the gorge, creating a snowpack that easily could have persisted into July and August and caused Randy to try to cross the creek farther upstream, at a smaller waterfall. Then again, he and some of the other rangers speculated, Randy might have fallen through a snow bridge and drowned and been sucked over the falls. Or he was traumatically injured at or somewhere upstream from where the radio lay, his remains ultimately washing down the creek. The radio and the backpack's waist belt, Durkee knew, were the most important clues thus far.

Backcountry ranger Kay Edens, who had taken over the Bench Lake ranger station after the original search was called off, proved her artistic prowess by sketching detailed drawings to help document the gorge and the locations of Randy's remains and gear. Edens began with a waterfall, which DeLaCruz and the rangers measured as being 150 feet upstream from the radio. Fifty feet downstream from the radio was another waterfall. It wasn't large by Sierra standards—dropping only 9 or 10 feet—but it was loud, even at low water. Directly above it the current was strong enough that a person could not stand. There the creek funneled into a narrow rapids, then poured into a chute that channeled the force of the water into a wedgelike spillway. The water was most powerful at that spot—the combination of vertical drop and narrow channel creating a brief but violent torrent that emptied into a large crack between slablike granite boulders. That was where Randy's body most likely had been pinned all these years.

The bright blue of his sleeping bag and the flutter of torn fabric in the rushing water had caught the eye of a searcher. The bag had unfurled and captured a number of items that were recorded in the order in which they were recovered from the creek. Interspersed between the items you'd expect to find in a ranger's backpack were individually identified bones. It was a morbid checklist: "NPS ball cap, lower mandible, MSR stove, bivy sack, scapula, comb, upper jaw, NPS jacket . . ."

Then something attached to a tan piece of fabric flashed in the water. Reaching in under the falls, a searcher carefully tugged at a familiar-looking National Park Service–issue shirt, with gold ranger badge, dented and corroded, still attached.

DeLaCruz let out an involuntary "Oh, jeez" when it became apparent there was something still in the shirt. With some pulling, the shirt came loose from where it had been wedged tightly between the rocks, revealing Randy Morgenson's name tag. George Durkee was there and made sure DeLaCruz saw that part of a clavicle was still inside: "critical," he says, "to prove Randy was in uniform and on patrol for the Department of Justice."

"The shirt made us pause," says Scott Wanek. "It was the one piece of evidence that really hit home. Even though we all knew what we'd found before and what we'd been collecting, seeing Randy's name put life into all these pieces that a few minutes before were kind of like inanimate objects. Now, all of a sudden, it felt disrespectful to touch anything."

Only the sounds of the creek and the falls could be heard; the rangers said nothing. Wanek broke the silence a few minutes later to radio Chief Ranger Debbie Bird. She had been waiting for confirmation so she could contact Judi Morgenson.

"It's Randy," said Wanek. "No doubt."

There was no gallows humor at this recovery operation, but speculation was voiced freely. The backpack had been found further downstream, along with various pieces of gear. It was theorized, matter-of-factly, that Randy's body had "come apart" here in the falls, which enabled the backpack to break away and be taken by the current downstream, where it and parts of his body were scavenged, as evidenced by bear and coyote tooth holes in a rusted tuna fish can Randy had been carrying.

Perhaps the owl had called Randy's name on that surreal ashen morning the season before his disppearance. If there was one thing Randy was tuned in to, it was the cycles of nature and the knowledge that he himself was part of those cycles. His decomposed body and scavenged bones certainly verified his thoughts along those lines. "Something to sing about," Randy had written after hearing the twilight howls of a pack of coyotes feasting on a deer carcass along the shores of Bullfrog Lake in the 1980s. Here in this remote gorge, sixteen bone fragments were all that remained of Randy Morgenson, glorifying the spirit of a man who said, "The least I owe these mountains is a body."

WHILE PACKING THE REMAINS into a rectangular Rubbermaid Action Packer, somebody unbuttoned the chest pocket of Randy's ranger shirt. Inside was a hand lens—something Randy had kept figuratively and literally close to his heart since 1980, the summer his father died. The lens had been Dana Morgenson's and was quite possibly the same one he'd used to show Randy the magnified worlds of alpine gold and sky pilot on Randy's first climb up Mount Dana when he was 8 years old. Dana had still carried it when he'd visited Randy at Tyndall Creek just before his death.

The rangers spread out and camped near the investigation site that night. The gorge was still peppered with yellow tape marking Randy's final resting place—500 linear and 50 vertical feet of pure, and now tragic, beauty. Randy could not have picked a more stunning place to die.

"The Sierra was at its best that dusk and dawn," says George Durkee, "and that gorge and surrounding cirque was about as pretty a place as I've ever been in the mountains. It was hard that night. The last few years before the search weren't the best between Randy and me, but that night and the next morning, I felt like I saw that place through Randy's eyes."

"Pure—untouched, untrammeled, unlettered . . . hard to even find a footprint," wrote Randy his first season as a backcountry ranger while hiking off-trail in an area similar to the Window Peak Lake basin. "High country—above treeline (not timberline out of respect for the trees) where grass and flowers grow in small patches or tufts between the boulders, small streams splash between grassy banks and gurgle under the boulders, and glacial tarns lie silent and rock bound, glistening in the sun. Rich country!"

"I missed my friend," says Durkee. "And out of respect, the sooner we got out of there, got all the tape and people and washed away the footprints, the better."

By morning, the rangers and investigators had hammered out the potential scenarios of Randy's death. Publicly, most agreed with Durkee's statement that "Randy couldn't have killed himself here if he tried," and even privately, Durkee would not be swayed from that opinion, despite the worry he'd felt during the search that Randy might have gone off to some special place to end his life. "The evidence told me otherwise," says Durkee, who, admittedly, wasn't the acid test for the suicide theory. Nor was he unbiased. He'd spent the better part of five years trying to pursuade the DOJ to grant Judi Morgenson the Public Safety Officers' Benefit, writing letters, corresponding with Judi, searching to find precedent-setting cases (there were none). At the time Randy was found, he was putting together a visual presentation to show that the terrain Randy patrolled "wasn't a walk in the park," that it was dangerous, remote, and could hide a body forever. If it was decided that "the death was caused by the intentional misconduct of the public safety officer or if the officer intended to bring about his or her own death . . . no benefit can be paid."

The morning after the recovery, Special Agent DeLaCruz carefully walked the area around the radio for 100 yards up- and downstream. He had to be meticulous because his report would inevitably be used to determine whether Randy had died in the line of duty or had taken his own life. With so few human remains recovered, the coroner would look to DeLaCruz for descriptions of the location where the body was found and the conclusions he'd drawn from investigating the site.

Throughout the course of the recovery, Durkee had offered DeLa-Cruz his opinion."With Randy's remains," says Durkee, "the DOJ couldn't possibly argue against, one, 'he'd died,' and two, 'in the line of duty.' " But Durkee didn't have much confidence in the DOJ "desk jockeys," whom he suspected "couldn't read a topographical map if their life depended on it."

DeLaCruz's opinion of the Sequoia and Kings Canyon backcountry rangers had changed since he'd first met them the year Randy disappeared. Their antics in training were still unlike anything he'd experienced in any other park, but they'd proven themselves a talented and proficient crew. He'd been particularly impressed with their mind-bending thoroughness the day before as they methodically investigated the gorge. "I couldn't have asked for a more talented group of men and women to assist in such a difficult investigation," DeLaCruz says. With the investigation completed, "There wasn't a doubt. It was an accident while Randy was on patrol."

At 9:30 A.M., a critical incident stress debriefing was held adjacent to the shores of Window Peak Lake. Debbie Bird had flown in to take part, and the rangers present said a few things about Randy, just as they had after the original search was called off. Someone had hung Randy's shirt from a tree branch, where it swayed in the winds sweeping up the canyon.

It was Sandy Graban who said, "Randy taught me how to appreciate *everything* about these mountains."

"Why does a flower, a tree, anything exist?" wrote Randy in 1966 while at Charlotte Lake. "Because the universe would not be complete without it."

That pretty well covered it.

JUDI MORGENSON had experienced more loss in the previous decade than she cared to dwell on. Her eldest brother passed away in 1992; her mother-in-law, Esther, in 1993; she heard about Randy's affair in 1994, at almost the same time that her mother was diagnosed with terminal cancer; her mother died in 1995; Randy disappeared in 1996; he was declared dead—without a body—in 1997, the same year another of her brothers passed away. In 1998, Randy's brother, Larry, died from the effects of alcoholism without knowing the fate of his brother. Now, on July 18, 2001, Judi was in the San Francisco Bay Area for the funeral of a good friend from high school. She was staying with Gail Ritchie Bobeda.

Chief Ranger Debbie Bird contacted Judi's brother, Bob Douglas, who gave her Gail's phone number. Frantic to talk to Judi before the news leaked to the media, Bird called her early in the morning the day after the recovery.

The moment she set the phone down, Judi began to cry, an involuntary reaction she had choked off in her throat too many times for too many years. "It was just a shock," she says. "The problem with missing people is it puts you in limbo forever. You just don't know. Even if you *know*, you just don't know. There had been so many cries in the past, but that one hurt the most. Right then it wasn't closure, it was just pain. I couldn't have been with a better person than Gail. I was so lucky to have been there with her when I got that call." It was Gail who had introduced them, who caught Randy in his affair, and who was there for this final phone call. Judi's relationship with Randy had "started with Gail and it ended with Gail," says Judi, "who has been my best friend for forty-five years."

Debbie Bird had kept the information to a minimum, stating that "they'd found some remains" they believed were Randy's, along with his radio and his shirt with his name tag and badge. "It's Randy," the chief ranger told Judi. "I wouldn't call you if I wasn't sure."

The question still remained: "How?"

Back in his closet-size Ash Mountain office, Special Agent Al DeLaCruz refreshed his memory from the original Morgenson SAR records and organized a second file titled "Recovery of Human Remains." The next day, he handed over Randy's remains to Fresno County deputy coroner Loralee Cervantes, who interviewed both DeLaCruz and George Durkee about Randy, the search, and the recovery operation.

The recovery of Randy's remains brought closure, but hardly put an end to the questions surrounding his death. Back at his duty station at Charlotte Lake, Durkee wrote Alden Nash:

So, when I awoke on Monday, another bright and bushy Sierra day, I did not expect to be holding Randy Morgenson's jaw in my hand that afternoon. Nonetheless . . . Strange days. As I'm sure you've heard by now, there's no question it was an accident, though we'll be darned if we can really figure out how. At first I thought a snow bridge collapsed, but just talked to one of the trail crew who went through there; he said there was no snow. Most likely Randy was crossing a small gorge/stream and slipped— either hitting his head or something. The water doesn't seem like it could be much, but it was obviously enough to cover him from dozens of searchers for five years. I expected much steeper terrain if we were to have found him.

Two weeks later, Bob Kenan returned to the investigation site, alone. He needed a clearer picture of the role he had played in 1996, and the ways Randy might have died. Some of his motivation was to discount the hovering specter of suicide. During the SAR, he hadn't been able to discount the possibility and had even wondered, somewhat shamefully, if Randy had taken the time to apologize to him at Simpson Meadow the season before the SAR in order to clear his conscience. "Personally, it took a hike back over to the area after the investigation before the time frame of when my group searched came back to me," wrote Kenan in his 2001 end-of-season report. "I remembered on this trip that it was the sixth day of the search . . . I was with a team that included Rick Sanger, Charlie Shelz, Dave Pettebone and Ned. We came from the north over Explorer Pass and camped at a small lake one-quarter mile north of the accident site. The next morning this 'A Team' combed the area down to Window Peak Lake, as we thoroughly searched the gorge. We actually walked feet away from Randy. The only way we would have missed him is that he was totally obscured from view underneath an ice pack. High water in this area would not have hidden him from our view."

Kenan considered the morning start a significant factor, and reiterated later how "fresh, focused, and rested" his search team had been. "There is NO WAY considering that we searched methodically down that drainage that we would have missed Randy.

"There had to have been snow. A lot of it."

Interpreting a simple note on his team's debrief would later validate this: "descended w. side of creek to Window Peak Lake." For years Kenan had *always* crossed over the creek from the west side to the east side when descending toward the lake. The fact that his search team stayed on the west side could mean only one thing: the creek had been too dangerous to cross. And the only way that gentle section of creek could be too dangerous to cross would be the presence of a great deal of snow.

On his second day at the site, Kenan awoke with a sense of clarity. Besides the "absolute certainty" that snow had been present in the gorge, he was 98 percent certain that Randy had hiked to this location from Bench Lake in one day (meaning the accident would have occurred on July 21, 1996, in the afternoon or early evening) and that the location of the radio also marked the approximate location of where Randy had died. The accident "must have occurred when Randy was crossing a snowfield over this drainage and he fell through," says Kenan. "He must have fractured a leg or something and been unable to pull himself out of the ice pack." Therefore, Randy had not died right away. It wasn't a peaceful passing. "He was in there and in the water," he says, "and he eventually died either from the injury or hypothermia."

But how did they miss seeing something? That still haunted Kenan, who went to the park records to clarify that his search team had indeed come through the area ten days after the date on which, he felt fairly certain, Randy had had his accident. After that length of time, concludes Kenan, "There would have been a hole in the snow bridge that, with the heat of the sun melting, would have erased any sign that someone had fallen through."

After reading his old reports, Kenan recalled two times since the search when he'd tried to come through the Window Peak Basin but was turned back by "freak snow storms" atop Explorer Col. He felt there was "no way" he would have missed seeing the radio at his standard creek crossing had he traveled that way in subsequent years. This revelation tuned him in to another possibility: "Maybe Randy wasn't ready to be found before now and the mountains honored that wish by hiding his body all these years."

With further research, Kenan was eventually informed that the forward-looking infrared (FLIR) helicopter that had flown the entire search area by night had been unable to fly the southernmost segments, including the Window Peak Lake area, due to high clouds on some of the ridges. The map that recorded the flight plan showed a large X directly over the Window Peak Basin. Had cosmic intervention sabotaged the search effort?

THE THEORY THAT RANDY had survived a snow-bridge collapse was contradicted by what one ranger called "the devil's advocates"—the switched-on radio and the buckled waist belt. However they looked at these two items of evidence, the rangers found them both telling and baffling.

If Randy had been seriously injured but conscious and in the water after the accident, the first thing he would have done, if he were able, was release his waist belt in order to rescue himself. If this had been the case—even if he'd died eventually from shock or exposure—his pack's waist belt would not have been buckled, unless he'd been so wedged into a spot in the ice he couldn't reach the belt. The buckled waist belt, therefore, appeared to support the theory that Randy had been knocked out or was, for some other reason, unconscious or already dead once he hit the creek under the ice and snow.

Randy usually had kept his radio in the top zippered compartment of his backpack or carried it in his hand. The fact that it was turned on and found separate from his backpack supported the theory that he was seriously injured but conscious and could not reach anybody for help. Just as likely, however, he could have been monitoring his radio or attempting to make a call when the accident occurred. If so, was he simply trying to make contact with park headquarters to check in? Or had he been in peril?

The public affairs officer for Sequoia and Kings Canyon, Kris Fister, had compiled all the press releases and newspaper articles from the time of the search. Among these papers was what appeared to be a press release written by Tom Tschohl, the acting chief ranger while Debbie Bird was in the backcountry with her family. The document, dated July 26, 1996, reads: "Dispatchers had a clear radio contact with Ranger Morgenson at the 1130 AM morning roundup [July 20, 1996]. The following day [July 21] the morning roundup reported a garbled message that was sketchy and unreadable. The message was believed to have been transmitted by Randy Morgenson. On July 22 and July 23 dispatchers were unable to reach Morgenson. On July 24, per protocol, a search was initiated by District Ranger Randy Coffman." That garbled message was so difficult to understand, it was eventually deemed inconclusive, and officially noted as not being from Randy.

All other official documents maintained that the last known contact had been July 20, when Morgenson was atop Mather Pass. Newspaper articles published at the time didn't mention the "garbled message," suggesting that this information was not made available to the media.

Some park personnel, primarily those who have not been to the site, say that the Window Peak drainage is in a radio "shaded" or "dead" area. This is not the case. The Window Peak drainage, including where the radio was found, has line-of-sight contact with the Mount Gould repeater, which implies that any technical problems would have been associated not with Randy's location but with the radio itself. Nobody could discount that very real possibility, especially considering the parks' history with unreliable radios in the backcountry.

Almost across the board, backcountry rangers had experienced issues with their radios that season. The Motorola MT1000 was new and used rechargeable batteries that lasted only two or three days. The old radio Randy had grown accustomed to had a battery life of about

six days. Bob Kenan speculates that Randy had been unfamiliar with the battery life on the MT1000 and left Bench Lake with an under-charged battery.

In 2001, when Randy's remains were discovered, Rick Sanger was working in an office, having taken the season off to help spearhead a computer programming project involving robotics.

When Sanger learned where Randy had been found, his detail-oriented mind began to analyze his memories; like the others, he was obsessed to learn the truth and to "understand how I might have failed." He agrees with Kenan that Randy probably had left his station with an undercharged radio, and he too couldn't remember if there had been snow in the drainage. For weeks Sanger scoured his notes, his log-books, and his photos, until he came across some photographs he'd taken two weeks before Randy's disappearance, including a "spooky" image of Laurie Church's trail crew at Woods Creek, where he'd met Randy to give him the radio and the last time he'd seen him. It was a classic group photo, the trail crew lined up holding the tools of their trade, their shovels. But when Sanger looked closely, he saw something he'd missed before. Randy was in the background, a dark silhouette under some trees—alone. The image was a haunting reminder of the emotional strain of the search.

Sanger also found a self-portrait from atop Mount Clarence King, taken in July 1996. In the background, over Sanger's shoulder, was the Window Peak drainage. Purely by luck, he had documented, in full color, that the basin was filled with late-season snow. Especially preva-lent in the photo was a long strip of white, which Sanger recognized as the creek and gully.

With the presence of snow verified and the suspicion of radio prob-lems generally conceded, Sanger ultimately agreed with the snow-bridge theory. He'd fallen through a small snow bridge once himself, and says it was "instantaneous and frightening." It doesn't take a very creative person "to imagine the horror you'd feel if you were in the middle of nowhere and compound-fractured a leg," says Sanger. "I've been there. I've seen snow bridges collapse, and they can be sudden and violent—the sheer force of falling through to one of the rocks in the creek would be enough if you were pinned and unable to stop the bleeding. . . . If shock doesn't consume you, hypothermia would, but not right away. Add to that a busted radio, and it's about the worst possible scenario I can imagine."

"I still wonder today if Randy would be alive if he'd had a working radio," says Lo Lyness. She, like Randy, had documented radio problems in her station logbooks throughout her career, including during the Morgenson SAR. "Communications are terrible," she wrote. "Cedar to Bench—what a joke. Radio shop has made lame attempts to improve the situation, nothing effective."

A number of safety recommendations for backcountry rangers came about as a result of Randy's disappearance, including an overhaul of the "morning round-up" system for tracking rangers; mapping the radio "dead zones" in the parks; investigating the purchase and use of personal locator devices; and the implementation of a mandatory system for rangers to convey ther patrol itineraries.

By 2001, the radio dead zones in the park had not been mapped, personal locator devices were not in use, and though satellite phones were being tested by administrators, the backcountry rangers were still using the same radios that Randy had deemed unreliable.

Lyness was informed that Randy's remains had been found upon her return from a trip into the Sierra backcountry. "I cried uncontrollably for at least ten minutes," she says, "and then continued to grieve for the next two months." For months to come, Lyness, like all the backcountry rangers, heard the theories about what had occurred in the Window Peak drainage as people tried to make sense of Randy's death. She ignored most of them. And she firmly denounced any mention of suicide. She had already considered the possibilities years earlier during the search and rejected the ones that didn't fit. Suicide, according to Lyness, "didn't fit."

"Randy wouldn't have done this to us," she says. "He just wouldn't. He'd have known *exactly* what would happen, he would have been able to visualize every step of the SAR process, he'd know how traumatized all his friends would be, that his *personal* stuff would be read, and that's not what he would have wanted. Randy may sometimes have been self-involved, but he wasn't cruel. To have gone off to commit suicide without leaving a note would have been deliberately cruel to those of us who had been his friends and colleagues. Plus, for heaven's sake, how would he have done it? He didn't have his revolver with him, he hadn't been out to the frontcountry to get quantities of drugs. I just couldn't see him slitting his wrists. It just didn't make any sense."

The one thing that did make sense to Lyness was Sandy Graban's theory that Randy had had some sort of medical emergency. "That's the only rational guess I've heard," Lyness says.

Complaining of chest pain, Randy had, in fact, seen a doctor the winter before his disappearance, but the doctor had attributed it to stress, not to any physical ailment. He was given a clean bill of health and deemed extremely fit. Nonetheless, if Randy had been experiencing a medical emergency such as a heart attack, that could explain why he would have been less than cautious while crossing the snow bridge.

Then there was the waist belt. In a medical emergency, would Randy have kept his pack on or taken it off? Logic dictates that Randy would have taken it off. Then he would have radioed for help. A massive heart attack or stroke that hit suddenly and decisively at the exact moment he was crossing the snow bridge, which then collapsed and hid his body from rescuers, seems far too coincidental, and thus improbable.

There was brief speculation that he'd been attacked by a mountain lion. "No way," says Kenan, remembering his own encounter years before. "In the extremely unlikely scenario that Randy saw a mountain lion, he would have sat down and had lunch with him."

DEPUTY CORONER LORALEE CERVANTES informed Special Agent Al DeLaCruz by telephone on July 31, 2001, that the human remains submitted to her had been positively identified via dental records as those of Randy Morgenson. Cervantes later received the opinion of a forensic anthropologist that "Mr. Morgenson's remains were scavenged, possibly by a bear, since all of the tooth marks and related damage to his skeleton are consistent with an animal of that size, and none of the damage was due to a pre-mortem injury. A large portion of his remains, however, are missing and probably were moved by the scavenger from the location where his body was discovered." In September, Cervantes's report, "In the Matter of Investigation Held Upon the Body of James Randall Morgenson, Deceased," was delivered to DeLaCruz, who quickly skimmed the document looking for the crux, which read:

> The exact circumstances of Ranger Morgenson's death are unknown and there are not sufficient skeletal remains to be able to determine with certainty the mechanism. . . . However, from investigative records and a description of the action and movement of the decedent prior to his death, as well as records of the search efforts, it is most probable that while on the job of patrolling his duty area as a Back Country Ranger for the National Park Service, Ranger Morgenson met with some unknown injury resulting in an accidental death at or near the creek in the Window Peak Area of Kings Canyon National Park on or about 7-22 or 7-23-1996 at an unknown hour.

This report set in motion a review of Judi's claim with the Department of Justice, and in record time, just a couple of months later, she received a flimsy brown envelope containing the check for $100,000 plus interest. Following five years of legal nightmares and more than a dozen instances when she nearly dropped the case, the check—without so much as a letter—was bittersweet. But there was vindication. Randy had been recognized, officially, for giving his life in the line of duty.

Both the coroner's report and DeLaCruz's account presented an unsettling detail about the SAR that might have saved Judi all those years of anguish. One of the search dogs had "reportedly alerted on a spot where the Ranger's remains were ultimately located," wrote Cervantes. DeLaCruz added in his report, "The dog was injured at the time and was taken out of the area." That dog, of course, was Seeker, Linda Lowry's giant schnauzer, which had fallen through the ice and nearly drowned just upstream from where Randy's remains were found. For whatever reason, no official inquiry was ordered to determine why Seeker's alert had not led to the discovery of Randy's body five years before. The question remains—like so many things about this incident—unanswered.

STUART SCOFIELD had gotten his hands on the investigation reports that described, in great detail, the terrain where Randy had met his end—and the theories of a snow-bridge collapse or a slip while crossing the creek. He didn't buy it. He *wanted* to believe that his friend had met with a tragic accident, but after going over the reports "with a fine-toothed comb," he remains convinced that suicide was the most probable explanation. Randy had shared some of his deepest, darkest feelings with Scofield, most of which he never revealed to anybody else. And something about Randy's ultimate demise didn't add up.

Scofield isn't the only one who believes this; he is just the only one willing to discuss it for the record. Even though the subject of suicide is generally taboo, he feels it needs to be put on the table. "Randy always spoke his mind," says Scofield. "Had he told me to keep my mouth shut, what I'm about to say would be in a vault. And if I'm wrong, if Randy did just have an accident, well, I hope nobody holds it against me for thinking this way.

"First of all," he says, "I can't deny the things he left for me—the Sermon on the Mount and all—that told me he knew he wasn't coming back after that season. Whether he was thinking about suicide or thinking about disappearing, there is no doubt he left those things as messages for me.

"You can't say that somebody didn't have an accident. I just don't believe Randy had *that* accident. I am a man of the mountains also, and, well . . . I want to make the analogy of a bicycle courier in New York City. They live a dangerous life, true. But there are certain accidents that they just are *not* going to have. It becomes second nature.

"After traveling for years in the backcountry, I swear to God you intuit whether the snow bridge is stable or not. It doesn't have anything to do with actual physical conditions and snow crystals and ice crystals and adhesion. You just know. And that is why I don't believe that *that* is an accident Randy would have had.

"Now, getting trapped by rockfall or something that was more catastrophic in nature, a pine tree blowing over while you're rafting down a river, an avalanche maybe—a completely random act. Not a mistake. Randy didn't make those mistakes, and that is just how I feel about it."

Indeed, Randy had spent a lot of time on late-season summer snow, suggesting that he knew where to travel. His logbooks are filled with passages documenting where he'd contemplated snow bridges, avoided them, sometimes hiked miles upstream to find a safer crossing. But he had also had a few accidents. He'd fallen on loose scree and broken his hand, for example. He slipped on log crossings more than once. Randy wasn't superhuman, but Scofield felt that his kindred relationship with the wilds was such that he wouldn't have made a fatal error. The cosmic nature—the beauty—of the spot where Randy was found, coupled with the essentially mellow terrain, leaves Scofield no other option than to settle on suicide, though he won't speculate about the actual mechanism.

If Randy had decided that suicide was his only recourse, it was speculated that he wouldn't have wanted his friends to know. He would have wanted his friends to believe that the mountains had claimed his life in their unfathomable and arbitrary way.

Another speculation was that Randy needlessly, and perhaps subconsciously, put himself in harm's way to tempt fate. George Durkee—staunch "accident" theorist that he is—confirms that Randy, more than anybody he knew, had a serious fatalistic attitude. "He wouldn't always wear a seat belt," says Durkee, "telling me once that 'When your number is up, it's up.' "

What better test of karma in wilderness than to tread across the frozen surface of a pond that may or may not hold your weight? But most of Randy's friends say, "That makes no sense." What does make sense, at least to three of his friends, is that Randy may have done it for Judi. One such friend, who knows the High Sierra intimately, figures that "Drowning in a frozen river or lake would have been the perfect way to go. He wouldn't have found some cliff to jump off, because, well, Judi wouldn't get the benefit. He would have assumed, if he'd gone missing on patrol, that people would presume he died in the line of duty." Those contradicting this idea say that Randy probably didn't even know about the Public Safety Officers' Benefit. The other speculated motivator was redemption for his guilt, knowing that Judi would be taken care of. Two facts back this suicide theory. One: Morgenson had made sure his divorce papers were not signed, sending them out of the backcountry just days before his disappearance. If he had signed them and agreed to the divorce, Judi would not have been an eligible recipient of the benefit. Two: Morgenson did not leave a note. As with all the theories, however, there are gaping holes.

The explanation for Randy's death given on the Officer Down Memorial Page, a Web site dedicated to law enforcement officers who die in the line of duty, posted: "Ranger 'Randy' Morgenson drowned after being swept over a waterfall while on a solo backcountry patrol in the Sequoia and Kings Canyon National Park, California." The statement invokes the image of a spectacular falls, a rushing torrent that had swept Randy off his feet and over the precipice. Not the gurgling creek that rarely gets more than a foot deep before cascading over a 9-foot drop.

This unintentional exaggeration confirms that Morgenson's death, like his disappearance, will always be open to speculation.

ON OCTOBER 13, 2001, nearly 200 people gathered at the Montecito Sequoia Lodge outside Kings Canyon National Park to say goodbye and to honor the life of backcountry ranger Randy Morgenson. Judi Morgenson had framed two dozen of Randy's photographs—predominantly mountain storms, desert vistas, craggy peaks—that were hung on the lodge's walls alongside a wildlife shot that Randy had taken when he was younger: a photo of a downy juvenile owl that had fallen from its nest, camouflaged among the ferns and undergrowth of a Yosemite forest.

Judi was in the front row of the lodge's main hall, sitting between her brother, Bob Douglas, and her best friend, Gail—both of whom had been by her side through the entire ordeal. Nearby was Randy's remaining family, a couple of sets of uncles and aunts and Randy's cousins. It was no mistake that George Durkee and most of the other backcountry rangers in attendance were in the very back row. In between were an assortment of park employees and friends, some in uniform, some not.

For hours, people recounted their memories. One of Randy's fellow backbencher ranger pals, Walt Hoffman, read from a list of passages taken from Randy's logbooks. Randy's childhood friend, Bill Taylor, had flown in from Seattle; Joe Evans, Randy's partner from his Nordic ski ranger days at Badger Pass, had come from Colorado. Both shared stories of how Randy helped them to keep their priorities straight—to take notice of their surroundings, to not rush through life, and to be gentle on the land.

Nina Weisman, sitting near the back, had realized she'd taken down Randy's Missing Ranger bulletin on the very same day his remains had been found. She couldn't deny that there were forces at work in these mountains she would never understand. Randy had once told her that if she was "quiet and still, the mountains would reveal their secrets."

Indeed they had—at least in part.

Rangers are, in general, a pretty stoic group, "but there were some tears that day," remembers Al DeLaCruz. Durkee, who had written and rewritten his eulogy for Randy a dozen times since he'd started it in his mind at the falls above Window Peak Lake, couldn't bring himself to read it at the memorial. It remained in his pocket on a folded piece of paper.

Toward the end, Debbie Bird stood and presented Judi with a plaque that she and the rangers had put together. Below an etching of the Sierra and the words "Sequoia and Kings Canyon National Parks" were Randy Morgenson's shined but dented badge and name tag. At the bottom of the plaque, two inscriptions. "Backcountry Ranger

1965–1996" and a line from Shakespeare's *Henry V*: "We few, we happy few, We band of brothers."

This "band of brothers," the backcountry rangers, took it upon themselves to name an "unnamed" peak in Kings Canyon after Randy. It's the first peak west of Mount Russell and just north of Mount Whitney—a high and wild granite monolith of 14,000 feet that was somehow overlooked all these years. From this day forward backcountry rangers who knew Randy or his legend began referring to it as Mount Morgenson. The name can't be found on a map—the U.S. Geological Survey and Sequoia and Kings Canyon National Parks don't officially recognize the title—but over time, it will stick. "You can't Google it," says one ranger, "but you can climb it."

And really, that's all that matters.

CHAPTER FIFTEEN

A MISSED CLUE

Ice, like other things in this world, only appears solid and immobile.

—*Randy Morgenson, Yosemite, 1978*

Neither fire nor wind, birth nor death can erase our good deeds.

—*Buddha*

ALDEN NASH DID NOT ATTEND the memorial, for no other reason than a bad cold that kept him in bed all day. He, like Lo Lyness, would pay his last respects somewhere in the backcountry.

For five years, Nash had conducted his own search for Randy's body. After Randy's body was found, whenever he met up with someone he'd supervised, the question about what had happened to Randy would mutate into a string of queries for rangers like Dario Malengo, George Durkee, and Bob Kenan, who were now the veterans of the tribe. "Tell me about this snow-bridge theory again" or "What was that gorge like? How deep did the water get?" As the questions stacked up in one such conversation, Durkee's wife, Paige, ultimately voiced her opinion: "Just let Randy die."

But Nash couldn't. He had been biding his time, listening to the

answers and painting his own mental picture of what the backcountry rangers had begun referring to as "the spot." Even without visiting
the gorge above Window Peak Lake, Nash was certain that Randy's
mental state had contributed to his death. He thought that Randy had
made a mistake—and paid for it dearly. But Nash could not completely
discount suicide, even though he trusted Kenan's and Durkee's snow-
bridge theories. He would have to go and see for himself, see what *he*
felt about the place.

JUDI MORGENSON STOOD on the back porch of her Sedona, Arizona,
home. With a glass of wine in hand, she gazed out over the lush green
river valley below that contrasted with the desert and red-rock mesas
dominating the landscape.

When they had first looked at this house, Randy had stepped
through the front door and without pause walked straight through the
entryway, past two bedrooms, a bathroom, the kitchen and living room,
and onto the back porch. He'd put his hands on the railing, scanned the
horizon, and without so much as turning on a water faucet or checking
the roof, said, "This will work."

The land was paramount. The land always had been.

Ten years after they'd moved to Sedona, in the summer of 2002,
Judi had company over for dinner, which began in the traditional
manner with wine, cheese, and crackers—not unlike the happy hours
the elder Morgensons had long ago hosted in Yosemite. How things
had changed. In Yosemite, that first date with Randy had been in the
art gallery where she'd shared with him her dream of someday exhibit-
ing her artwork in just such a place. Today, Judi Morgenson's ceramics
are found on the shelves of the finest galleries in Sedona. Each piece,
she says, "has a little bit of Randy in it—the Sierra, nature. I can tell
you, Orange County isn't a hotbed of inspiration, at least it wasn't for
me. Randy took me away from all that."

During a brief tour of her home, Judi told her guests how Randy's
boxes still hadn't been unpacked. She pointed them out, many sitting
where George Durkee and Paige Meier had originally stacked them

in the attic, guest bedroom, and basement. Judi tended to stick to the rest of the house, feeling "squeezed by the memories" of her late husband. "It's not like I'm saving it for our kids," she said, but she couldn't bring herself to go through all of it. For Judi, not having children was one major regret in her life. She remembered the conversation she'd had with Randy before they were married, how they agreed they didn't want children. Randy had used overpopulation as part of his reasoning, "but really, he didn't want to be tied down," said Judi. "He didn't want anything to threaten his summer job or change his lifestyle."

It was another example of how Randy could not reconcile his life in the mountains with ordinary life outside that world. For the first half of their marriage, Judi had enjoyed that freedom as well, "but later on, I did regret it," she said. "I told Randy that if I had been married to a different man, who wanted children, I would have."

Over a bottle of wine, a sunset, and a meal, she spoke of her late husband. If nostalgia counts for forgiveness, it seemed that Judi had forgiven Randy his mistakes in life. She seemed determined to preserve his memory as the kind wilderness protector, and would just as soon have forgotten about the "other stuff."

Even though Judi had been deeply hurt by Randy's actions on more than one occasion, she understood that his most heated love affair hadn't been with some "other woman," or even with her—it was with the High Sierra. In a sense, when she finally decided to make the separation permanent, the divorce papers had been walking papers. She had simply set him free.

For a long time after his disappearance, Judi had avoided reading Randy's journals because it was too painful. She had, however, read his last personal journal, after it was returned to her by Al DeLaCruz. The DOJ profiler had used it as the basis for designating Randy an "elevated suicide risk," and it had included the stream-of-consciousness style of writing he'd learned from *The Artist's Way*.

Judi will not disclose the contents of that journal to anyone. It's "the only thing I have from Randy that's mine and nobody else can have it," she says. "I'm going to keep it for me."

ALDEN NASH TOOK two days getting to the Bench Lake ranger station—the Taboose Pass Trail in August "can be a bitch." It was 2003,
and he thought he'd finally heard enough conjecture to follow his own
gut and retrace Randy's last patrol.

There was no Bench Lake ranger station that summer. The budget
hadn't allowed for it, and any volunteer manning the station was now
gone. All that remained in the gravel among the lodgepole pines were
a picnic table, some bear boxes, and the wooden tent platform. After
sitting at the table for a spell with his hiking companion, Nash walked
back toward the trail. In his way was a perfectly formed obsidian arrowhead perched atop the loose soil mound from a Belding's ground
squirrel's burrow. Picking it up, Nash said, "That's Randy testing us."
He pushed the arrowhead deep into the soil with his heel, keeping it
from someone's pocket for a few more years.

He turned left on the Taboose Pass Trail, crossed the creek, headed
north on the John Muir Trail and onto the Bench Lake Trail. He paused
respectfully at an empty campsite, a flat spot under the evergreens
and amid classic Sierra boulders. This was the location from where he
and Randy Coffman had carried the body of the 17-year-old girl who had
died in 1991.

Farther down the trail, a decades-old, sun-bleached, weathered
board nailed to the trunk of a tree made Nash pause. "I bet if Randy
were here right now," he said, "he'd tell us exactly what that was for."

Randy *had* come upon such a board on July 13, 1996, between Saw-
mill Pass and Woods Creek—just one week before he disappeared. He
wrote in his logbook: "I found a camp with a board nailed to a tree
trunk where in the '60s we stapled cardboard 'Mountain Manners'
signs. Dick McClaren era. How many working today would see that
board and understand?"

A few more yards and Nash bent to pick up a granola bar wrapper,
stuffing it in his pocket. "Another test," he said with a wink.

Throughout the course of his career, by the numbers recorded in a
portion of his available logbooks and EOS reports, Randy had collected

some 600 gunnysacks full of "backpacker detritus." Predominantly full of glass and cans, each sack weighed around 35 pounds—21,000 pounds of garbage that Randy had removed from the backcountry.

At the end of the Bench Lake Trail, Nash crossed into the muted, pine needle–muffled world of the real backcountry—the route he knew Randy had taken. At this point, if Randy had veered right instead of left, he would have ended up over Cartridge Pass and ultimately in Lake Basin. Here, at this fateful Y intersection, Nash recalled Durkee's statement about how he and the other rangers had been "gloriously wrong" in the area where they'd focused the search.

Probably the most frustrating thing during the SAR was that Randy had not left behind an itinerary. Randy felt that spontaneity was a big part of the wilderness experience, and in fact he preached its benefits. When Sequoia and Kings Canyon were considering the implementation of a wilderness permit system in 1971, Randy wrote to the Sierra District Office: "One virtue of the wilderness experience . . . is the unstructured, unplanned, relatively spontaneous mood . . . something which will be lost if we initiate a reservation controlled use system."

Therefore, it's conceivable, even likely, that Randy did not leave an itinerary at Bench Lake because he himself had only a vague idea of where he would be going. That's a decision he probably made when he was met with the intersection of the Taboose Pass, John Muir, and Bench Lake Trails, choosing the latter—a dead end, but to Randy, a doorway. Carrying four days of supplies, his only clear intention was that he would be going cross-country. This freedom and lack of a defined route would have made Randy feel happy, or at least normal, during that turbulent time.

"All of your life, someone is pointing the way, directing you this way and that, determining for you which road is best traveled," he wrote in his 1973 McClure Meadow logbook. "Here is your chance to find your own way. Don't ask me how to get to McGee Canyon or Lake Double-Eleven-0. Go, on your own. Be adventuresome. Don't forever seek the easiest way. Take the way you find. Don't demand trail signs and sturdy bridges. Don't demand we show you the mountains. Seek them and find them yourself. . . . This is your birthright as an animal, most commonly denied you. Be free enough from intentions to find goodness wherever you are and in whatever is happening. Here for once in your life you needn't do anything, be anywhere at a determined time, walk in a certain direction. You can now live by whim.

"Here's your one chance to get lost, fall in the creek, find a beautiful place."

Nash took most of the day getting to a high, lonely, and "indescribably beautiful" campsite he felt Randy might have chosen himself, then carefully spread his sleeping bag on a flat section of gravel between two granite boulders, folding his ground cloth to avoid a tuft of meadow grass that "was trying to make a living." Looking to the west that evening, he watched a dramatic sunset through the haze of a distant forest fire. Everything was bathed in an orange glow, bats were sweeping insects off the bubbling creek, and Nash said aloud, "How do you explain this place?"

The following afternoon, Nash made it over the Arrow Peak ridge via a "wicked-dangerous" nonroute. He had barely paused for an afternoon "nutrition break" before he started comparing the rocks at his feet to the rocks in a photograph he'd obtained from Durkee, which showed exactly where Randy's radio had been found. He was standing in the gully above the gorge, at "the spot" in the Window Peak Lake drainage.

It was the first week in August, and there was still a snowbank 3 feet deep along the shaded eastern wall of the gully. Walking up and down the stream, examining the high-water marks on the rocks and estimating the water at 6 inches deep, Nash finally perched himself on some rocks at the top of the falls and shook his head. "It's a bloody mystery," he said. "It makes sense that Randy was crossing up there where the radio was found; there's nothing between there and the falls that would stop a body. I'll have to compare snow-depth histories."

Back at the radio's location for the fifth time in an hour, Nash said, "I don't know how he would have done it, but I guess he could have offed himself right here."

At that moment, thunder rumbled. Looking up at Pyramid Peak to where a thunderhead had sneaked into the basin, he suggested finding a place to camp. By the following morning, he had decided that "there must have been a lot of snow—Bob and George are right about that." Without snow, there was no possible way Randy could have met his end in "that benign little creek.

"Impossible."

Nash continued his cross-country loop from Bench Lake to the John Muir Trail, back over Pinchot Pass; eventually, he camped near the Bench Lake ranger station site, where he gathered his thoughts while waiting out another late-afternoon thunderstorm under a poncho. Finally he had a good mental picture of "the spot," but he was still chewing on the options, spitting them out as he went. He leaned toward the snow-bridge theory, but felt something was missing. *He* was missing something.

A couple of weeks later, a friend who'd taken an interest in the "Morgenson saga" called Nash at his home in Bishop to inform him that he'd reviewed the search records and found that a search dog had expressed interest farther upstream from where the radio had been found. The dog had been injured falling through a frozen lake and wasn't able to continue the search.

That got Nash's wheels spinning.

"Tell me what you know," he said. "Tell me about this search dog."

ON JULY 30, 1996, the seventh day of the search for Randy Morgenson, Linda Lowry and her injured dog, Seeker, were being flown from Window Peak Lake to Cedar Grove. As the helicopter carried them across the mountains, Lowry became more and more convinced that Seeker had been onto something. Seeker was a talented search dog with a good track record. She had never been known to take Lowry "on a walk," search-dog slang for "a wild-goose chase."

At Cedar Grove, Lowry was debriefed at 9 P.M. Her most pertinent responses included: "Dog showed interest and tried to track at 4084.5N/370.4E. But her feet were injured, and there had been other hikers there recently [referring to Dave Gordon and Laurie Church, who she was told had searched the segment five days earlier]. Not enough time to search more extensively, needed to evacuate dog." For "Difficulties?" Lowry confirmed that snow was heavy in the area, even for late July: "Snowfields close to lake (dog fell in at one point). Loose talus and scree." For "Suggestions?" she recommended, "Would take another dog back to pt. where Seeker showed interest, and work all the way to PCT from Window Peak Lake."

It was a straightforward suggestion. Lowry figured that the incident command team would read it and send another dog back to the immediate vicinity where Seeker had fallen through the ice, where she had taken the time to mark the GPS coordinates. If two dogs showed interest, or alerted, there would be no doubt that Seeker had been on to something. The next dog would tell the tale, she thought.

At the end of the next day, Lowry saw Eloise Anderson at the incident command post. Anderson had been somewhat of a mentor to Lowry as she and Seeker had gone through the CARDA certification process. During their conversation, Lowry found out that it was Anderson who had been flown back to the Window Peak drainage to take over where Seeker left off. Anderson said she hadn't been informed of Lowry's recommendation. She certainly wasn't given the GPS coordinates that Lowry had recorded.

Had there been an error? Was Lowry's recommendation overlooked by the incident command team in the stress of the SAR? Or was it considered and—like the Window Peak area itself—deemed a low priority? That could be the reason why the next search dog continued where Seeker and Lowry had been picked up by the helicopter, on the shores of Window Peak Lake, instead of going back to cover the area Lowry had recommended, just a quarter mile upstream.

It was in that quarter-mile gap that Randy's remains were found. Nobody can remember with certainty why Lowry's recommendation hadn't been followed. Scott Wanek, who was the operations chief during the original SAR, took part in the recovery, and attended both of Randy's memorial services, had been at the planning meetings each day of the search. He remembers that they were meticulous, but, he explains, "searching for people is not an exact science. It will always boil down to making judgment calls about a very complex array of human behavior, terrain, and weather. We can be fairly scientific about organizing resources and keeping track of where we have searched, but ultimately, just about every critical decision during a search has to be made by people based on human judgment using information available at the time." For whatever reason, the information at the time told the overhead planning team not to send Anderson's dog to the location Lowry had recommended.

Was it possible, then, that Lowry's recommendation had been ignored or overlooked—had there been a mistake?

"If the information was simply lost in the shuffle somehow, then I would consider that a mistake," says Wanek. "But if it was obtained and evaluated and the decision was made to give it a low priority for follow-up, then I definitely would not consider it a mistake." Wanek's gut feeling is that the information was evaluated and given a low priority.

Debbie Bird, who calls the search for Randy the most difficult SAR she has ever been involved with, says, "There was never any doubt in my mind that we had covered all the bases, some a number of times—everything was done that could have been done." She does admit, however, that "the one thing I would have done differently would have been to bring in a fresh incident command team, probably composed of people from outside Sequoia and Kings Canyon, midway through the search in order to bring fresher, rested minds to the search. But by the time it really occurred to me, it was time to start talking about scaling back the search."

In June of 1997, Bird organized a formal board of review—made up of search-and-rescue professionals from other parks such as Glacier, Cascades Cluster, and Yosemite—that critically evaluated the Morgenson SAR. The board concluded that the search had been "conducted in accordance with policy and using the best available technical knowledge and equipment." However, "While command staff was not necessarily formally trained to be qualified to occupy the Incident Command Positions as filled, the experience and technical knowledge levels present in the IC Team provided for incident management at a level sufficient to insure that the best possible search effort was made."

Still, a number of problems were identified regarding the SAR. They included the parks' "inadequate" radio system and how "Routine tracking of backcountry staff (round-up) error resulted in delay before search initiation. The somewhat lax method of daily accountability for the backcountry staff resulted in a probably 24-hour delay. . . ." Most of these problems were discounted in the board's official report, with such disclaimers as "None of the communications issues adversely impacted the outcome of the search and some of the problems have subsequently been corrected."

Bird maintains that her reason for requesting the review in 1997 had been to critique her staff's performance and to potentially "learn from our mistakes." The discovery of Randy's remains in 2001 represented the true "outcome of the search," yet no board of review has critiqued why Lowry's suggestion had not been acted upon—what might have been a major mistake.

Randy's remains were found approximately 150 feet downstream from where Lowry had recommended the incident command team follow up. Of course, Randy was certainly dead long before Lowry and Seeker entered the drainage. In the end, it was, at the very least, frustratingly bad luck that they had come so close yet had not found him then. It would have eliminated five years of anguish for Judi and Randy's friends and colleagues, and his body would potentially have told a more definitive story.

ONCE ALDEN NASH HEARD the story about the search dog, he began to think in terms of snow and ice. He speculated that whatever had happened to Randy might have occurred farther upstream, near where Seeker almost drowned.

At the recovery investigation, search dogs and searchers had not found any of Randy's remains or equipment upstream from the radio, and they logically concluded that his death likely occurred at or very near the spot where the heavy Motorola radio had embedded itself in the creek bed.

Nash also learned that Lowry, upon seeing photos of the gully and gorge, was "astounded by the depth of the gully along the creek." She had remembered it as all snow, with little or no indication of a gully. That being the case, Nash reasoned, there would have been up to 10 or 15 feet of snow in places along the creek, no doubt left over from winter avalanche activity. This also explained with virtual certainty why Kenan's search party had not crossed the creek: it had been too dangerous. But ten days earlier, before a storm cycle brought rain to the area, it would have appeared more stable and crossable. That was when Randy had been in the drainage.

Nash talked with ski guides who travel in the high country during the winter. He consulted the California Department of Water Resources Cooperative Snow Surveys data for both Charlotte Lake and Bench Lake—the nearest snow survey courses to the Window Peak drainage. He used his contacts to obtain satellite images of the area, and George Durkee then mapped the GPS coordinates from Lowry and compared them to the location where the radio had been found. At first there seemed to be a small problem: Lowry's GPS coordinates placed Seeker and her in a cliffy area on the western slope of the drainage—far above the creek and nowhere near any water. Knowing a dog couldn't have fallen through ice where there was no water, Durkee deduced that on the GPS coordinates, the correct number, 9, had been recorded as a 4 on the debriefing form, likely occurring when Lowry read her handwritten scrawl from the field to the debriefer at Cedar Grove. Once Durkee plugged the corrected number into his computer mapping program, the location was right on the edge of a small but occasionally deep pool that, according to Durkee's program, placed Seeker 97 feet upstream from where the radio was located. Using these new coordinates, Nash zoomed in on the satellite images and, from his memory of "the spot," formulated an entirely different theory:

Randy had not fallen through a snow bridge at the location of the radio—rather, he'd attempted to cross the creek farther upstream from west to east and, like Seeker, had fallen through the ice. But when Randy crossed the area ten days before Seeker did, it likely looked like solid snow, which he would have assumed concealed a narrow creek beneath—not the deep water of a hidden pond. "If Randy went through the ice," says Nash, "the marks would have disappeared in a day or two because of the rapid melting and settling of the snowpack. In a very few days it would look like the usual winter snow cover or lake ice as it softened during the day and refroze at night. It would have taken an unusually astute observer to sort that scenario out in those circumstances, and maybe with the dynamics of that snowpack it would have been totally unsortable."

"However," says Nash, "the dog would know."

According to this scenario, "It would have been over quickly and solves the waist-belt mystery. The fact that it was attached on his backpack suggests something quick and disastrous, medical or otherwise." Lowry remembers Seeker "swimming against a current" when she'd fallen through the ice, enlarging the hole with her paws, which illustrated how thin the ice was at the time. Ten days earlier, the ice was likely topped by more snow and would have been thicker. If Randy had punched through, he would have dropped deep into the freezing water because of the weight of his backpack. The current would have pushed him immediately away from the hole in the ice. After a few frantic attempts to resurface, his hands would have been numb from the freezing water, making it nearly impossible to unbuckle his waist belt. He would have been trapped with his backpack on, under the ice, for little more that 30 seconds before panic, hypothermia, and lack of oxygen ushered in death.

"At that elevation Randy would have been preserved in place for quite some time," says Nash. "Probably until very late in the season and possibly frozen in place until the next winter." Nash explains how the two years following Randy's disappearance had been above-average snow years. Avalanches would or could have buried the small pool in question with even more snow. "That is way wicked steep country above that creek, and it slides constantly in winter and spring," he says.

But how did Randy's body end up 150 feet downstream from where he'd fallen through the ice without leaving behind any remains or gear? Nash answers that question by telling the story of a World War II–era plane that crashed in a training flight on the Mendel Glacier in the north end of Sequoia and Kings Canyon. The plane did not surface until the 1960s, when pieces of debris and occupants began appearing on the retreating face of the glacier. "These mountains can swallow you up and spit you out when and where they decide," says Nash. "I have observed more than once the evidence of a large snow avalanche crashing down on a lake and scooping all or part of the lake out and completely over the outlet side of the lake. Everything—fish, water, logs, and rocks—ends up in a debris field below the lake. In our small lake in question, this would put everything in the channel just above where Randy's radio was lying. As I recall, the channel was a classic debris field of stuff that had migrated there from above through natural erosion and snow and rock slides. This pool was already in the channel, so the spring high water would wash the debris in question down into the gorge where Randy was found."

Nash's theory explains why Kenan, Sanger, Gordon, and the other searchers had not seen Randy near the area where the radio was found—at the time of the search, he would have been under snow and ice farther upstream.

Regarding the switched-on radio: "It is a normal situation for rangers to be walking along with the radio on and zippered into the top pocket of their pack with just the antenna protruding out while monitoring—or in anticipation of the morning roundup," says Nash. "In this case and with the lake-ice theory, the radio might not have left the pack for years." Nash explains how Randy's body would have been "rolled" down the creek with his backpack on, and the antenna would have slowly worked the zipper open, or "some critter" had tried to get at Randy's lunch that would probably have been on top, next to the radio. Eventually, the radio slipped out of that pocket right where they found it, and "it was heavy enough to stay put while Randy and his pack continued to roll downstream another 50 feet—probably being nibbled on along the way—and didn't stop until he was wedged in those falls."

Nash seems to have a well-thought-out response for all the major points of contention—the waist belt, the movement of the body to its final resting point, the radio being switched on. But what about Randy being there in the first place? Certainly Randy's time in the mountains had tipped him off that he was crossing a potentially dangerous area, whether it was a snow bridge over the creek or a snow- and ice-covered pond.

"First off, you can't ignore Randy's state of mind at the time," says Nash. "That alone put him at a major disadvantage. There is no doubt in my mind that Randy's mind-set had something to do with his death." That said, "In the Park Service, we're often asked to identify the hazards in a certain area. But how does a backcountry ranger alone on patrol identify hazards when the whole damn place is a hazard? Any ranger route is a hazard, and in between one hazard and another hazard is *another* hazard—like a snow bridge. You've got to get across a creek, and chances are it's stable. You've done it many times before. Sometimes you've gone around it, but usually you've walked as lightly as possible and done it. But it's hard to walk softly with a 50-pound pack. It's roulette, that's what it is. I guess after twenty-eight years of good luck, you're gonna hit black when you bet on red."

Nash's theory—that Seeker fell through the ice on the same pond where Randy had drowned beneath its frozen surface—even holds something for those with an appetite for the supernatural. First, there was Judi's dream of the man with a backpack at the bottom of a lake. Then there was the "psychic" backpacker who told rangers that she'd had a disturbing vision of a man trapped underneath something. Finally, in Robert Bly's *Iron John*, the book that Randy had read so avidly, the bearded "wild man" in the story is discovered at the bottom of a pond or small lake when a dog is pulled into the water. This from the book that Randy had said "spoke to me."

PERHAPS RANDY was having a glorious day in the Sierra when he met his demise. Flowers were blooming along the edges of snowbanks and new grasses were sprouting up. Birds were singing, ground squirrels foraging, marmots lazing in the sunshine, pikas chirping. If so, he may have been lulled into a sense of bliss. For twenty-eight seasons he'd strolled through these mountains—across chasms, along lakes, over snowfields—thousands of times, making such observations as the one he recorded in 1973, when he watched a "small band of rosy finches chattering quietly with their deep voices while running and jumping across gravel and bare sod, between clumps of short haired sedges and grasses, harvesting seeds off the sedge-tops. Watching without disturbing, for these mountaineers aren't readily disturbed, a feeling of goodness about the world comes over me. If things are well for the rosy finches, what ill can befall me?"

For the span of his career, Randy sensed there were messages coming to him while in the mountains. That inspired summer in McClure Meadow in 1973, he wrote, "I am suddenly close to something very great and very large, something containing me and all this around me, something I only dimly perceive, and understand not at all.

"Perhaps if I am here, aware, and perceptive, long enough I will."

We can never know for certain what occurred in the Window Peak drainage. Compelling arguments all, but can there ever be absolute closure without speculation? The absence of a definitive answer seems appropriate, considering Randy's love of mystery in the mountains.

In his files Randy kept an Albert Einstein quote that his father had loved. His mother had included it in a memorial she wrote for the *Yosemite Sentinel* when Dana died in 1980:

The most beautiful thing we can experience is the mysterious. It is the source of all true art and science. He to whom the emotion is a stranger, who can no longer pause to wonder and stand wrapped in awe, is as good as dead; his eyes are closed. The insight into the mystery of life, coupled though it be with fear, has also given rise to religion. To know what is impenetrable to us really exists, manifesting itself as the highest wisdom and the most radiant beauty, which our dull faculties can comprehend only in their most primitive forms—this knowledge, this feeling is at the center of religiousness.

And in Randy's own words, from a logbook dated September 12, 1978: "How can I claim to a greater importance than these alpine flowers, than anything that lives here, or even than the very rocks which eventually become the nourishing soil from which it all has to start? The existence of souls in men? And who can tell me the souls do not take up residence in plants and animals, or even these waters and rocky peaks? A higher evolution for the souls in men? So does that make us more important? Everything has its place, everything supports everything else, everything is important to itself—to its own development— and to that which it supports.

"That a humanoid God willed all this into existence simply to glorify himself (a bit too egotistically human), and/or for us, his greatest creation, and our pleasure, use, misuse, seems not either to fit with the way I perceive the world while living close to it here at Little Five Lakes.

"I wish only to be alive and to experience this living to the fullest. To feel deeply about my days, to feel the goodness of life and the beauty of my world, this is my preference.

"I am human and experience the emotions of humanity: elation, frustration, loneliness, love. And the greatest of these is love, love for the world and its creatures, love for life. It comes easily here. I have loved a thousand mountain meadows and alpine peaks.

"To be thoroughly aware each day that I'm alive, to be deeply sensitive to the world I inhabit and the world that I am, not to roam roughshod over the broad surface of this planet for achievement but to know where I step, and to tread lightly.

"I would rather my footsteps never be seen, and the sound of my voice be heard only by those near, and never echo, than leave in my wake the fame of those whom we commonly call great."

May your trails be crooked, winding, lonesome,
dangerous, leading to the most amazing view . . . where
storms come and go as lightning clangs upon the high
crags where something strange and more beautiful
and more full of wonder than your deepest dreams
waits for you . . . beyond the next turning of the canyon
walls.

—*Edward Abbey, "Benediction"*

I can't decide whether I want to spend my next life as a
little alpine bird or as a marmot. We should be careful
before concluding that either of these would be stepping
down.

—*Randy Morgenson, Rae Lakes, 1965*

IN MAY 2003, George Durkee, in full-dress National Park Service
uniform, accompanied Judi Morgenson to Washington, D.C., where
Randy was honored by having his name added to the National Law
Enforcement Officers Memorial on Judiciary Square. Judi ceremoni-
ously placed a rose for Randy on a wreath representing all the peace
officers who had died in the line of duty that year.

After the ceremony, at a reception held by the Department of Interior, Durkee honored his friend by making public the eulogy he'd writ-

ten, but could not bring himself to read, at Randy's memorial service a
year and a half earlier:

> We have come together to tell each other stories about our friend
> Randy and so try to bind him more firmly to our memories and
> our lives. In a too often chaotic universe, it is our shared memo-
> ries that will help bring a sense of order—a common narrative—
> to a life lost. Like everyone else here, and I think especially Judi
> and the backcountry rangers, I've been telling myself a variety
> of stories over the last five years and none of them made much
> sense; none brought any peace. Although it reopened old and
> painful wounds for all of us, finding Randy at last gives us a way
> to heal and helps to answer the most painful question of this
> story—that there was nothing we could have done.
>
> And so I'll tell the story I've begun for myself. Today and
> in the coming years others can add theirs. From our collective
> memories we begin to weave a tapestry of a life that will keep
> him with us.
>
> Wherever he is, I don't envision Randy's spirit smiling
> beatifically down on us from amongst a heavenly host. Nor is he a
> warm and fuzzy pika chirping at us from among alpine boulders.
> There was a fierce and wild energy to him—a misanthropy
> that kept him independent of others. Years ago several of the
> backcountry rangers, assigning totem animal spirits to each
> other (we don't have cable out there . . .), decided Randy was
> a wolverine—probably the ultimate symbol of wildness in
> the Sierra. For me, and especially in the last five years, he's a
> raven, riding uneasily on my left shoulder and looking out at
> the world with his unblinking brown eyes, muttering thoughts
> and opinions; occasionally pecking at my ear to draw attention
> to what's around me; even occasionally drawing blood. Of all
> of us in the backcountry, Randy's vision was the clearest, his
> wilderness philosophy the purest. He was—and for me still is—
> the conscience of the backcountry.

He's left us small windows into his vision. The photographs are his enduring legacy, his most sustained attempt to bring that vision to the rest of us. We have lyrical passages from his station logbooks reminding us to pay attention. And there are lessons I take with me from Randy's life and death—some of them not easy. The first is the most obvious: be careful out there. If the best of us can fall in what struck us as easy terrain, that's a clear warning to spend the time to look for an easier crossing; to study an area a little more closely before moving across it; to take longer naps.

The other lessons that keep rattling around in my brain are a little harder. Randy was struggling with himself his last few years. He and I were not always easy on each other then; his pain radiated out to all of his friends at one time or another. We do not come with an owner's manual to help us through such times nor, as friends, a blueprint to offer help so the offer is heard.

Randy's last hike brought him to a narrow gorge in a high and remote alpine basin. An ancient stream rushes down that gorge and, though always facing the open sky, its roots lie in arctic twilight. From the cliff walls come the questioning cries of rock wrens. Distantly the ethereal call of a hermit thrush measures shadows moving slowly across the canyon. Then darkness and a murmuring stream, running down and over rocks, spray flying toward distant stars. Out into the stillness of an alpine lake, plunging yet again down and down, merging into the steady roar of the Kings River. Again, swiftly, down in a wild torrent past mile-high cliffs and sleeping trees leaning over steep banks, dreaming of warm spring days and bears rubbing against bark.

Finally, flowing quietly out into the great plains of the Central Valley, stars and a profound night sky take him. From the first mindful drip of melting snow to immanent silence is one continuous and joyful Sierra chorus. Randy's voice has joined that song and, listening quietly, we will always hear it.

AUTHOR'S NOTE
AND
ACKNOWLEDGMENTS

During the course of researching this book, various people—mostly rangers—asked me how I came upon this story and whether I knew Randy Morgenson personally.

Unfortunately, no, I didn't know Randy, though I wish I had. I began backpacking in the High Sierra when I was fourteen; ten years later, in 1992, I hiked the John Muir Trail solo through Sequoia and Kings Canyon at the suggestion of Dan Dustin, an inspirational professor of outdoor recreation at San Diego State University. In preparation for that hike, my dear friends Kathy and Craig Cupp introduced me to Alden Nash, who was nearing retirement as the Sierra Crest subdistrict ranger. Alden invited me into his home in Bishop, spread a map out on his dining room table, and provided me with invaluable insight on the JMT—including a suggestion: "Stop by the McClure Meadow ranger station and say hello to Randy Morgenson." Alden said that taking a walk with Randy would be like taking a walk with John Muir himself.

When I passed through McClure Meadow, I found a note on Randy's cabin door that read "Ranger on patrol, back this evening." Being a trail pounder on a schedule, I moved right on through and never had the chance to meet Ranger Randy. If I hadn't been in such a rush, I'd be recounting that meeting right now.

The following year, I joined Alden and Craig for the first of what would become annual August or September sojourns into the High Country. Inspired by these trips, I pitched an article to *National Geographic* that I titled "The Backcountry Rangers of the John Muir Trail."

At the time, I was a journalism student with one published article under my belt. Needless to say, my query was turned down.

In 1996, we hiked in over Bishop Pass after Alden informed us that Randy Morgenson had gone missing and that we would be on the lookout for his body. Randy "may have gotten into some trouble in wicked country," he told us. The route he chose was exceptionally gnarly; we traveled cross-country for seven days to ultimately exit the mountains at Taboose Pass. It was just weeks after the official SAR had ended; Randy's Overdue Hiker bulletin was still posted at trailheads.

Each summer thereafter, Craig and I joined Alden as he continued his search for Randy's remains, and as Alden revealed bits and pieces of "Morgensonia," I took notes with the faint notion of writing a magazine article, maybe even a book. After three years of covering "Randy routes," we hadn't found a single clue, though we often scrutinized skeletal remains that always proved to be those of an animal. In 1999, Alden lent me a copy of Randy's 1965 Rae Lakes logbook. He called it an "artifact" and joked that it was protected by the Antiquities Act, his way of saying that he wanted it back after I was done.

The entry that captivated me the most was read at both of Randy's memorial services and is found on page 269 of this book. The emotions and descriptions Randy conveyed during his first season as a backcountry ranger were so pure and genuine, I could only imagine what prose he had created during ensuing years, when age and experience had strengthened his bond with the mountains.

The writer in me wanted badly to read the other twenty-eight years of logbooks that Alden told me were scattered "all over the park," in backcountry ranger stations, buried in file cabinets and drawers of desks, piled in closets doomed for Dumpsters, and filed away in the Sequoia and Kings Canyon museum archives. The rest, he told me, could be found in Randy's attic in Sedona, Arizona, "guarded by his widow."

By this time, Alden had introduced me to a handful of rangers who encouraged my interest in the "Morgenson Mystery" by adding to Alden's storytelling and rattling the skeletons in Randy's proverbial closet. I was told of the affair and the pending divorce and that Judi Morgenson had been fighting the Department of Justice for denied death benefits. I was made privy to various theories regarding Randy's disappearance: suicide, a new life in Mexico, the morbid idea that an irate backpacker had murdered Randy and packed his body out of the mountains in pieces. The possibility of an alien abduction was relayed to me quite seriously.

All this, though very intriguing, wasn't exactly an inviting topic to discuss with Randy's closest friends and family and especially Judi Morgenson, who was the link to accurately portraying Randy's life story. At this point, Judi hadn't been afforded even mildly comforting closure. I surmised that she certainly wouldn't want to talk to a complete stranger about her missing, presumed-dead husband.

On July 15, 2001, I received an e-mail message from Alden that consisted of three words: "They found him." I knew immediately who "him" was.

I waited four months after the second memorial service to introduce myself and the idea of this book to Judi via e-mail. She thanked me for my interest and brushed me off with kind regards. I didn't follow up for another five months. Without Judi's blessing for this project, I told myself, I'd just walk away from it.

Almost a year after Randy was found, Judi called me and asked about my experiences in the Sierra. I babbled on for nearly an hour before she said, "You're already halfway there." I asked her what she meant, and she explained how knowing the Sierra was half the battle of understanding Randy. For the other half, she invited me and my wife, Lorien, into her home in Sedona, where she told us that Randy's writings, records, photographs, and belongings had "squeezed" her between the attic and basement for "far too long." For three days, she allowed me to sift through boxes and bookshelves and let me into the darkest recesses of her memory while Lorien camped out in front of a photocopier. We left Judi with little more than my promise to tell the story accurately. Her wish was that I convey the magic and mystery of the Sierra that called Randy back into the mountains for twenty-eight seasons. She also knew that I could not tell the story without telling the *entire* story.

Three years ago, I joined Alden Nash and retraced what he felt were Randy's final steps from the Bench Lake ranger station to "the spot" in the gorge above Window Peak Lake. I realized then, after chasing Randy's ghost for more than eight years, that the telling of his story began and ended with Alden. I can't claim to know what it was like walking with Randy, or John Muir, but I imagine either would be something like the hundreds of miles I've hiked with Alden, who sat on the rocks above the falls where Randy's body was found and shook his head and said, "What a beautiful place to die."

Shortly after I returned home from that trip, George Durkee called to tell me that Randy's name had also been added to the Peace Officer Memorial in Tulare County, where Sequoia and Kings Canyon are located. With chagrin, he said that Randy had officially been recognized for his service more in death than in life. He also told me that the National Park Service still has no nationwide program to award seasonal park employees—backcountry rangers or otherwise—for length of service.

I submit Randy's story to those in a position to implement such a program, so that others with his dedication don't slip through the cracks.

THIS BOOK WOULD NOT HAVE BEEN POSSIBLE without the unlimited, candid, and brave contributions of Judi Morgenson, Alden Nash, and George Durkee. I simply cannot thank them enough.

Other brave souls who were critical in the telling of this story include Dave Ashe, Debbie Bird, Tina Bowman, Al DeLaCruz, Sandy Graban, Lo Lyness, Doug Mantle, Cindy Purcell, Stuart Scofield, Barbara Sholle, William Taylor, Carrie Vernon, and Scott Wanek. Nina Weisman, Bob Kenan, and Rick Sanger hosted me in the backcountry for multiple days. Two years after Randy's disappearance, Rick was awarded the highest honor for a public safety officer—the Medal of Valor—by President Clinton for his role in the rescue of a drowning man not far from the Window Peak drainage. Also notable is Bob Kenan's recently produced documentary of life on the John Muir Trail. You can learn about it via his Web site: www.messagefromthemountains.net. Ward Eldredge in the Sequoia and Kings Canyon museum/archives department and Bob Wilson in the SEKI law enforcement department deserve special mention, as they did everything short of allowing me to set up camp in their offices, putting up with my constant requests for documents, photographs, and historical records. David Kessler of the Bancroft Library at UC Berkeley made a daunting task less daunting.

Robin Ingraham's emotional interview in which he recounted the death of his best friend awed me. As a result of Mark Hoffman's death, Robin eventually hung up his climbing gear and now photographs the Sierra with large-format cameras. His gallery on Yosemite and Sequoia and Kings Canyon (www.robiningraham.com) will take you visually to the heart of the land you've just read about.

Dozens of National Park Service personnel (some retired, some still in service) contributed time, memories, and guidance through uncharted territory in the search for archives and data. Some were included in the narrative, but many were not. Many thanks to Peter Allen, Cameron Aveson, Gail Bennet, Paul Berkowitz, Debbie Brenchley, John Dill, Kay Edens, Joe Evans, Butch Farabee, Greg Fauth, Kris Fister, David Graber, Terry Gustafson, Sylvia Haultain, Walt Hoffman, Ned Kelleher, Steve Klump, John Kraushaar, Ralph Kumano, Mark Magnuson, Rachel Mazur, Jeff McFarland, Bob Meadows, Paige Meier, Bob Mihan, Eric Morey, Jeff Ohlfs, Chris Pearson, Paige Ritterbusch, Tim Simonds, Rick Smith, Peter Stephens, Jerry Torres, Bill Tweed, and Scott Williams.

Darren Chrisman, April Conway, and Erick Studenicka of the Nevada National Guard helped me portray a critical component of the search-and-rescue operation, as did Linda Lowry with her thorough explanation of a search dog's role in a SAR. Thanks as well to the Mariposa County High School records department for the history lesson.

My wonderful agent, Christy Fletcher, who was by my side from this idea's conception, put the proposal into the hands of a number of editors who believed in the project. HarperCollins and editor Mark Bryant—the former editor of *Outside* magazine who had worked with such writers as Jon Krakauer and Sebastian Junger—were the perfect fit. Mark has a superb knack for guiding writers down their own storytelling paths. I was disappointed when he moved on from Harper-Collins, but he handed the baton to seasoned editor Henry Ferris, who continued to invoke my confidence as a writer and saw the manuscript through to completion without so much as a speed bump. He and his assistant, Peter Hubbard, fielded my calls and caffeine-induced e-mails with style, grace, and always a cheerful attitude.

Many thanks to Henry Arnebold, Liza Bolitzer, Laurel Boyers, Tim Brazier, Melissa Chinchillo, Chris Cosgriff, Bob Douglas, Linda Eade, Mark Fleishman, Emily McDonald, Marilyn Meyer, Jane and Jim Morgenson, Marc Muench, Sue Munson, Bruce Nichols, Dale Oftedal, Holly Russel, Randy Rust, Kate Scherler, Norma Snelling, Peter Stekel, Beth Sullivan, Pat Wight, and Nancy Williams-Swenson. Frederick Ludwig of Procopio, Cory, Hargreaves & Savitch guided me through the maze of intellectual property rights; Page Stegner granted permission for the lengthy excerpts from his father's letters; and Esther Shafran and the Ansel Adams Publishing Rights Trust allowed me the use of letter excerpts.

My support group of colleagues and friends helped me see the forest through the trees—or in some cases, the trees through the forest. Thank you: Matt Baglio, Alison Berkley, Kevin Blakeborough, Andy Blumberg, Lee Crane, Aaron Feldman, Tom Gonzalez, Scooter Leonard, Richard Leversee, Derek Mathis, Marcelle and Pete McAfee, Lisa Miscione, Joan Nash, Stephanie Pearson, Norman Peck, Torey Piro, and Dean Zack.

A number of writers whom I respect very much found time in their schedules to read advance copies of the manuscript. Thanks go to Greg Child, Daniel Duane, Butch Farabee, Nora Gallagher, Jennifer Jordan, Amy Irvine McHarg, Bill McKibben, Aron Ralston, Gene Rose, Julian Rubinstein, and Jordan Fisher Smith. Photographer Bill Hatcher also volunteered his time to read.

Very special thanks are in order for:

My on-call editor and muse, Randolph Wright, and tireless copy editor Rita Samols, both of whom donated long, long hours critically reading the manuscript at different stages in order to save me from myself.

My father, Clayton Blehm, who has shown enthusiasm for my endeavors without prejudice my entire life and is likely my biggest fan. My mother, Jacqueline Blehm, who passed away when I was seventeen and who has continued to guide me with her eternal advice: "If there is something you want to do, do it today, because you don't know about tomorrow."

My entire family and family-in-law for their unconditional support, wisdom, and for accepting "the book" as my excuse for nearly all of my shortcomings and canceled plans over the past few years: Beth Cloud, Nick Cloud, Debbie and John Cloud, Lori and Rick Hennessy, Shannon Hennessy, Amber Warner, Evan Warner, Heidi and Jeff Warner, and Judy and Fred Warner. Additional thanks to my brother Steve Blehm, who tediously tallied logbooks, and to my niece and nephew Madison and Mitchell Tybroski, who photocopied mountains of research material.

My escape to Neverland: Merrick, who spent the first part of his life being spoon-fed "Morgensonia." Perhaps this has instilled some of the love for nature that Randy had. And my incredibly supportive wife, Lorien Warner—my constant sounding board, my closest confidante, and my most critical editor, who painstakingly scrutinized every word.

Last, but not least, I must acknowledge the book's main character— the High Sierra. Thank you for calling out to me.

—Eric Blehm, September 2005

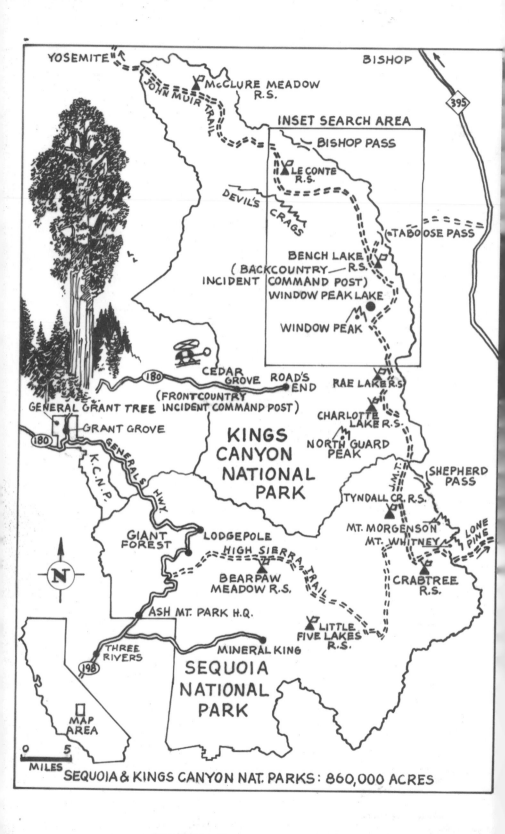

YOSEMITE

BISHOP

395

McCLURE MEADOW R.S.

JOHN MUIR TRAIL

INSET SEARCH AREA

BISHOP PASS

LE CONTE R.S.

DEVIL'S CRAGS

TABOOSE PASS

BENCH LAKE R.S.
(BACKCOUNTRY INCIDENT COMMAND POST)

WINDOW PEAK LAKE

WINDOW PEAK

CEDAR GROVE ROAD'S END
(FRONTCOUNTRY INCIDENT COMMAND POST)

180

RAE LAKE R.S.

CHARLOTTE LAKE R.S.

NORTH GUARD PEAK

GENERAL GRANT TREE

180

GRANT GROVE

K.C.N.P.

GENERALS HWY.

KINGS CANYON NATIONAL PARK

SHEPHERD PASS

TYNDALL CR. R.S.

MT. MORGENSON
MT. WHITNEY

LONE PINE

GIANT FOREST

LODGEPOLE

HIGH SIERRA TRAIL

BEARPAW MEADOW R.S.

CRABTREE R.S.

ASH MT. PARK H.Q.

LITTLE FIVE LAKES R.S.

N

THREE RIVERS

198

MINERAL KING

SEQUOIA NATIONAL PARK

MAP AREA

0   5
MILES

SEQUOIA & KINGS CANYON NAT. PARKS: 860,000 ACRES